Grid Computing
in
Life Sciences

LSGRID2005

The 2nd International Life Science Grid Workshop

Grid Computing in Life Sciences

Biopolis, Singapore 5 – 6 May 2005

Editors

Tan Tin Wee

National University of Singapore, Singapore

Peter Arzberger

University of California, San Diego, USA

Akihiko Konagaya

RIKEN Genomic Sciences Center, Japan

World Scientific

NEW JERSEY • LONDON • SINGAPORE • BEIJING • SHANGHAI • HONG KONG • TAIPEI • CHENNAI

Published by

World Scientific Publishing Co. Pte. Ltd.
5 Toh Tuck Link, Singapore 596224
USA office: 27 Warren Street, Suite 401-402, Hackensack, NJ 07601
UK office: 57 Shelton Street, Covent Garden, London WC2H 9HE

British Library Cataloguing-in-Publication Data
A catalogue record for this book is available from the British Library.

GRID COMPUTING IN LIFE SCIENCES
Proceedings of the 2nd International Workshop on Life Science Grid, LSGRID 2005

ISBN 981-270-378-0

Printed in Singapore by World Scientific Printers (S) Pte Ltd

PREFACE

The inaugural International Workshop on Life Science Grid (LSGRID 2004) was held in Kanazawa (Japan) on 31 May – 1 June 2004. Riding on the overwhelming success of the first workshop, LSGRID 2005 was organized by the Agency for Science, Technology And Research (A*STAR), Association for Medical and BioInformatics Singapore (AMBIS), Asia Pacific Bioinformatics Network (APBioNet), National Grid Office (NGO) and National University of Singapore (NUS) on 5 – 6 May 2005, as a part of GridAsia 2005 conference in Singapore. It received kind sponsorship from the Initiative for Parallel Bioinformatics (iPAB) of Japan. The second workshop was held back-to-back to the PRAGMA 8 Meeting.

Like LSGRID 2004, this year's workshop focused on life science applications of Grid systems especially for bio-network research and systems biology which require heterogeneous data integration from genome to phenome, mathematical modeling and simulation from molecular to population levels, and high performance computing.

The two keynote addresses at LSGRID 2005 were:

> "Grid as a 'Ba' for Biomedical Knowledge Creation" by Prof. Akihiko Konagaya (RIKEN Genomic Sciences Center, Japan).
>
> "CIBs: Cyber-Infrastructure for Biosciences" by Prof. John Wooley (Associate Vice Chancellor for Research, University of California San Diego).

The two invited addresses were:

> "Does Grid Technology Help Life Sciences? Lessons Learnt from BioGrid Project in Japan" by Prof. Shinji Shimojo (Director, Applied Information System Division, Cybermedia Center, Osaka University, Japan).
>
> "Grid Usability Case Studies: Deployment of Bioinformatics Applications" by Dr. Wilfred Li (Executive Director, National Biomedical Computation Resource, USA).

A panel discussion on "Life Sciences Grid Standardization" was chaired by Associate Professor Tan Tin Wee (National University of Singapore), with Prof. John Wooley and Dr. Wilfred Li as panelists.

Besides the keynote and invited speeches as well as panel discussion, some 20 oral presentations were made. This volume of post proceedings comprises the revised versions of the accepted papers of the LSGRID 2005 workshop.

We gratefully acknowledge the contributions of all the Program Committee members in reviewing the submitted papers. Thanks also go out to the authors and participants of the workshop for their oral presentations, revised papers, discussions as well as exchange of knowledge and experience. In particular, we thank Dr Lee Hing Yan, Chairman of the Local Organizing Committee and his team at the National Grid Office for his kind coordination and tremendous assistance in making the event a success and this book publication possible.

We look forward to meet you again at future LSGRID workshops.

Tan Tin Wee

Peter Arzberger

Akihiko Konagaya

Editors
August 2006

ORGANIZING COMMITTEES

Local Organizing Committee

Chair	Lee, Hing-Yan (National Grid Office)
Members	Ang, Larry (Bioinformatics Institute)
	Bhalla, Vineta (Ministry of Health)
	Choo, Thong Tiong (National Grid Office)
	Lau, Khee-Erng Jon (National Grid Office)
	Lim, Beng Siong (Singapore Institute of Manufacturing Technology)
	Tammi, Martti (National University of Singapore)
	Tan, Tin Wee (National University of Singapore)
	Yates, Brian (Blueprint Asia Pte Ltd)

Program Committee

Co-Chairs	Arzberger, Peter (UCSD, USA)
	Tan, Tin Wee (NUS, Singapore)
Members	Akiyama, Yutaka (AIST CBRC, Japan)
	Ang, Larry (BII, Singapore)
	Angulo, David (DePaul Univ., USA)
	Bala, Piotr N. (Copernicus Univ., Poland)
	Kao, Cheng-Yao (NTU, Taiwan)
	Konagaya, Akihiko (RIKEN GSC, Japan)
	Konishi, Fumikazu (RIKEN GSC, Japan)
	Lin, Fang-Pang (NCHC, Taiwan)
	Luo, Jingchu (CBI, Peking Univ., China)
	Matsuda, Hideo (Osaka Univ., Japan)
	Matsuoka, Satoshi (TITECH, Japan)
	Mohamed, Rahmah (UKM, Malaysia)
	Moss, David (Birkbeck College, London Univ., UK)
	Napis, Suhaimi (UPM, Malaysia)
	Rodrigo, Allen (Univ. of Auckland, New Zealand)
	Satou, Kenji (JAIST, Japan)
	See, Simon (SUN Microsystems, Singapore)
	Sekiguchi, Satoshi (AIST GTRC, Japan)
	Shimojo, Shinji (Osaka Univ., Japan)
	Sinnott, Richard (National eScience Centre, Glasgow, UK)
	Stevens, Rick (ANL, USA)
	Wooley, John (UCSD, USA)

CONTENTS

THE GRID AS A "Ba" FOR BIOMEDICAL KNOWLEDGE CREATION

AKIHIKO KONAGAYA

Advanced Genome Information Technology Research Group
RIKEN GSC
1-7-22, Suehiro-cho, Tsurumi, Yokohama, Kanagawa, Japan
Email: konagaya@gsc.riken.jp

Data-driven biology, typified by the Human Genome Project, has produced an enormous amount of experimental data, including data on genomes, transcriptomes, proteomes, interactomes, and phenomes. The next pursuit in genome science concentrates on elucidating relations among the data. Key to understanding these relations are bio-networks that incorporate information on protein-protein interactions, metabolic pathways, signal transduction pathways, and gene regulatory networks. Development of these bio-networks, however, requires biomedical knowledge of life phenomena in order to add biological interpretations. In this sense, data creation and knowledge creation play complementary roles in the development of genome science. As for knowledge creation, Ikujiro Nonaka proposed the importance of "ba", that is, a time and place in which people share knowledge and work together as a community. Grid computing offers great potential to extend the concept of "ba" to networks, especially in terms of deepening the understanding and use of bio-networks by means of sharing explicit knowledge represented by ontology, mathematical simulation models and bioinformatics workflows.

1. INTRODUCTION

In April 2003, HUGO proudly announced the completion of the Human Genome Project [1]. The success of that project has opened doors for further post-genome sequence projects that produce genome-wide data at multiple levels, for example, on transcriptomes, proteomes, metabolomes, and phenomes, to name a few. The next challenge is to elucidate bio-networks that incorporate information on protein-protein interactions, metabolic pathways, signal transduction pathways, gene regulatory networks, and so on. To understand these bio-networks, it is necessary to introduce biological interpretations to molecular-molecular interactions and pathways. Scientific experts provide such interpretations implicitly. However, the explicit representation of biological knowledge through such structures as ontologies and mathematical simulation models is necessary in order to be able to analyze bio-networks from a computational point of view.

Knowledge creation requires a time and place in which people share knowledge and work together as a community. Ikujiro Nonaka called this place "ba" [2], as originally proposed by the Japanese philosopher Kitaro Nishida [3]. "Ba" can be considered a type of superstructure similar to a virtual organization or community based on mutual trust. This paper discusses how to organize a "ba" for grid computing [4] from the viewpoint of biomedical knowledge creation, that is, for developing and interpreting computational bio-networks.

Section 2 of this paper describes the differences between data, information and knowledge using gene annotation as an example. Section 3 discusses the role of tacit knowledge in bioinformatics. Section 4 introduces knowledge-intensive approaches to drug-drug interaction based on ontology and mathematical modeling. Finally, Section 5 discusses the superstructure of grid computing necessary for creating and sharing biomedical knowledge.

2. DATA, INFORMATION AND KNOWLEDGE

Although the boundaries between data, information and knowledge are somewhat unclear, each differs from the viewpoint of "interpretation". "Data" is self-descriptive in nature. We can transfer data from person to person without explanation, for example, the nucleotide sequence "atcg". "Information" is objective in the sense that its interpretation is almost unique. Given a nucleotide sequence, you may find a coding region using the gene structure information together with information on the translation initiation and termination. On the other hand, knowledge is subjective in the sense that its interpretation depends on the individual's background. For this reason, we need ontologies and mathematical models to represent "descriptive" knowledge that is shared or is supposed to be shared like information.

Genome annotation is a knowledge-intensive bioinformatics application that maps the structural and functional information of genes. To annotate a gene, in-depth understanding of the functionality of the gene is required, by integrating information from genome, transcriptome, proteome and other databases. Terms play an essential role in genome annotation. Imagine that we are given homologous sequences by BLAST search for a specific coding region. If the sequences are all ESTs (expressed sequence tags) or unannotated genes, our understanding of the coding regions is limited. However, when we are given terms such as 'biotin carboxylase' and 'campylobacter' in the annotation of the sequences, we obtain more knowledge associated with the terms. The terms may remind biochemical experts of proteins and pathways related to the carbon dioxide fixation reaction. They may also remind medical or pharmaceutical experts of diseases, for example, cheilitis caused by the deficiency of biotin carboxylase and enteritis caused by campylobacter [5,6]. This example suggests two important aspects about terms. First, terms play a role in linking knowledge to other knowledge. Second, the semantics of a term depend completely on the expertise and knowledge of the scientist. We will discuss the characteristics of personal knowledge from the viewpoint of knowledge creation in the next section.

3. TACIT KNOWLEDGE, EXPLICIT KNOWLEDGE AND KNOWLEDGE SPIRAL

Michael Polanyi, a 20th-century philosopher, commented in his book, *The Tacit Dimension*, that we should start from the fact that *'we can know more than we can tell'*. This phrase implies that computers are limited in their ability to represent knowledge, no matter how fast they can calculate and no matter how much storage. they may have. Furthermore, in his book *The Knowledge-creating Company,* Ikujiro Nonaka observed that the strength of Japanese companies does not result simply from the solid management of explicit knowledge but from the establishment of common tacit knowledge. This does not, however, indicate the superiority of tacit knowledge over explicit knowledge. Explicit knowledge is important for analyzing and interpreting huge data sets such as genome sequences. To clarify this issue, Ikujiro Nonaka developed the concept of the "knowledge spiral," which turns tacit knowledge into explicit knowledge (externalization) and explicit knowledge into tacit knowledge (internalization), as shown in Figure 1.

Consider how the knowledge spiral could be applied to bioinformatics applications. A gene ontology has been developed to control the terminology used for genome annotation by strictly defining the relationship of terms [7]. From the viewpoint of knowledge spiral theory, genome annotation can be considered a type of knowledge transfer from tacit knowledge to explicit knowledge (externalization). Gene ontology also serves to label gene clusters through gene annotations (combination). These annotations then help biologists understand the functionality of genes (internalization). Finally, the process can be extended so that everyone in the community can share the same understanding of gene functionality (socialization). In this way, we can create new explicit knowledge (the annotation of genes) and tacit knowledge (an understanding of gene functionality) by repeating the above process throughout the community.

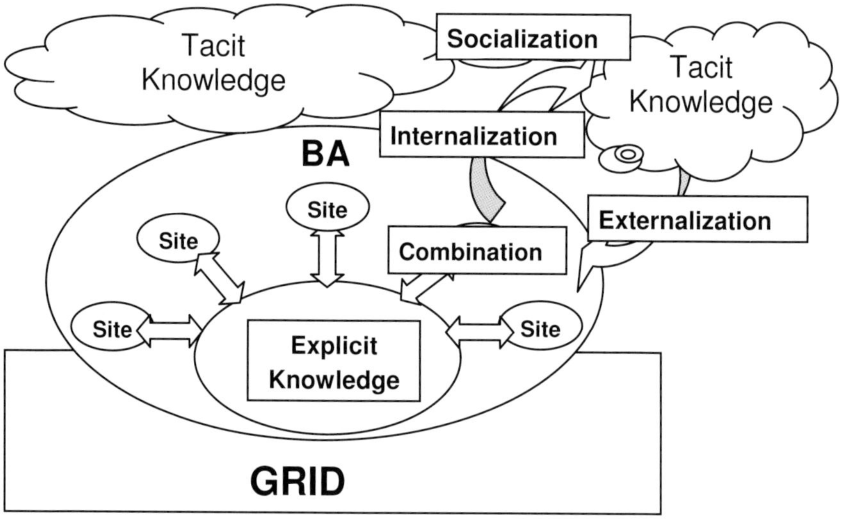

Figure 1. Knowledge Spiral on a Grid

Mathematical modeling of biological phenomena is another interesting application of the knowledge spiral to bioinformatics. Mathematical models model the dynamic behavior of biological phenomena, that is, the time-dependent state transition of molecular interactions. To design mathematical models, information on biochemical reactions and kinetic parameters is needed. Many efforts have been made to extract this information from literature databases [8, 9]. However, the information in the literature is fragmented and sometimes contradictory in terms of bionetwork modeling. Consider, for example, protein-protein bindings and mRNA expression profiles. Mathematical modeling also requires in-depth biological knowledge and strong mathematical skills in order to integrate information obtained from biological experiments. The development of mathematical models is still a state-of-the-art process performed by human experts knowledgeable in the life phenomena of interest [10]. Mathematical models, once established, however, can be extended to gene disruption models and over-expression models. They can assist our understanding of phenomena more deeply, helping us better understand the many efforts involved in biological experimentation. In this way, new explicit and tacit knowledge can be created.

4. KNOWLEDGE-INTENSIVE APPROACH TO DRUG-DRUG INTERACTION

Drug-drug interaction is a significant clinical problem. It was recently recognized that drug response strongly depends on the polymorphism of drug response genes such as cytochrome P450 (CYPs) [11]. Severe drug side effects may occur when a patient is a poor or extensive metabolizer of a given drug. This problem becomes more complicated when more than two drugs are concomitantly administered. To address this issue, we developed a drug-drug interaction prediction system based on a drug interaction ontology and a stochastic particle simulation system that incorporate drug-molecule interaction information extracted from the literature and sequence databases. The system will incorporate personal genome information in the near future.

When designing a drug-interaction ontology (DIO), we focused on a triadic relation consisting of *input*, *effector* and *output* [12]. The triadic relation represents the causality of molecular-molecular interaction in a cell. *Input* indicates a trigger of this molecular interaction. *Effector* indicates the central player in this molecular interaction. *Output* indicates the result of this molecular interaction. *Input*, *effector* and *output* for drug metabolism consist of a drug, an enzyme and a drug metabolite. Note that an *output* can be used as an *input* for a succeeding triadic relation. In this way, we are able to represent a drug metabolic pathway as a chain of triadic relations. Drug-drug interaction can be represented as triadic relations that share the same *input* or *effector* in their metabolic pathways.

A triadic relation can be extended to incorporate an indirect molecular interaction such as a metabolic pathway as well as a direct molecular reaction such as an enzymatic reaction. In other words, our system is able to represent metabolic pathways as a single triadic relation by ignoring intermediate reactions. The system is also able to represent the causality of a high-level molecular reaction, such as the inactivation of enzymatic function and the inactivation of drug function in cases for which biological observation is available but the molecular mechanism is unknown.

To date, we have extracted more than 3,000 interactions from the literature and entered these into the molecular interaction knowledge base. We have also developed a prototype system that infers the occurrence of drug-drug interaction in triadic relations.

A triadic relation is sufficiently powerful to represent drug-biomolecular interactions qualitatively, but is limited in its ability to analyze the dynamic behavior of quantitative information. Drug metabolism is highly non-linear, and drug response sometimes becomes sensitive to the initial drug dosage. This situation becomes more complex when more than two drugs are concomitantly used, as shown in Figure 2. In the figure, 6-mercaptopurine (6MP) is an anti-cancer drug, and allopurinol is an anti-hyperuricemia drug used to reduce the purine bodies that often result from cancer therapy. It is well known, however, that allopurinol inactivates xanthine oxidase (XO), which metabolizes 6MP to thiourea, which ultimately is excreted in urine. Severe side effects may occur in patients unable to excrete 6MP [13].

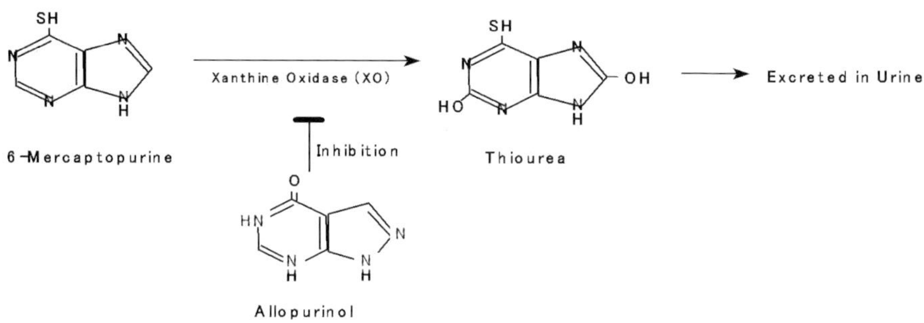

Figure 2. Drug-Drug Interaction between 6MP and Allopurinol

Although an ordinal difference equation system is the leading approach to analyzing drug metabolism processes, we adopted a spatio-temporal stochastic particle simulation to analyze trafficking processes, localizations, and membrane penetration [14]. The particle simulation system simulates a molecule as a particle that walks randomly on a 2D or 3D lattice grid unless boundary conditions, such as a membrane, are provided. Each molecular interaction may occur with a specified probability when reactive particles pass the same node of the grid. Our particle simulation is sound in the sense that the average behavior of a kinetic reaction is the same as that obtained from ordinal rate equations when the number of particles is sufficient and the particles are distributed uniformly. Our particle simulation can also account for the non-uniform distribution of particles and the behavior of particular molecules such as DNA, membrane structures, and receptor complexes. The particle simulation has good potential to overcome the limitations of conventional simulations that are based on ordinal differential equation systems, partial differential equation systems, and other deterministic simulation systems.

5. SUPERSTRUCTURE FOR BIOMEDICAL KNOWLEDGE CREATION

The concept of the grid computing has great potential to accelerate knowledge creation by extending knowledge spirally throughout the network community. Knowledge spiraling requires a platform upon which people share knowledge and work together. A grid enables users to share data, information and knowledge as well as computational resources. It should also emphasize the social aspects, or the superstructures constructed on the IT platform, that play an essential role in collaborative works in virtual organizations [15, 16]. We will discuss these issues from the viewpoints of community formulation, service interoperability and intellectual property development.

As described in the introduction, development of community is a basis of "ba" for knowledge creation. Grid users with network accounts establish a community to share data, software and computers over the network. In the case of a grid that uses Globus Tool Kit, the boundary of the community is restricted by the availability of Globus accounts to access the remote computers. The boundary can be relaxed if a single representative login account is provided for access to the remote computers as shown in the Open Bioinformatics Environment (OBIEnv) [17]. OBIEnv enables a local user to access the remote computers thorough the representative account if the local user has a valid account on a local machine. The community can thus be extended to the total number of local accounts on the grid.

The use of a reverse proxy server is another approach to extend the community. The reverse proxy server can provide the portal accounts needed to control access to the necessary web pages. The portal accounts enable the extension of the community to non-computing professionals, such as experimental biologists who prefer to access data and tools through web pages. As of July 2005, more than 400 portal users had registered for the bioinformatics applications on the Open Bioinformatics Grid (OBIGrid[i]). One successful example is the genome annotation support system (OBITco) developed for the *Thermus thermophilus* research community in Japan [18].

Grid services and workflows enable portal users to automate bioinformatics activities performed using hands-on web applications. Most public databases and well-known freeware bioinformatics applications are already available as web services[ii]. Knowledge management, that is, the sharing and reproduction of bioinformatics workflows, is a key challenge for knowledge creation.

---

[i] http://www.obigrid.org/

[ii] http://industry.ebi.ac.uk/soaplab/ServiceSets.html

Interoperability is important to ensure proper data transfer among applications. XML formats enable input and output data to be described in an architecturally independent manner. However, bioinformatics workflows require interoperability of semantics as well as data format. Let us consider a simple bioinformatics workflow for the annotation of a microbial genome: Glimmer2 [19] for gene finding, BLAST [20] for homology search, and CLUSTAL W [21] for multiple alignment. This workflow seems reasonable, but would produce unsatisfactory results if the commands were consecutively executed. This is because Glimmer2 may produce too many gene candidates. BLAST may return very similar but unimportant sequences such as EST (expressed sequence tag) or annotated sequences. We therefore require filtering processes that are able to eliminate redundant and irrelevant data from the computational results [22]. A bioinformatics service ontology is therefore needed in order for the community to make use of and share bioinformatics workflows.

Intellectual property on a grid is another important issue to be resolved from the viewpoint of knowledge creation. Who owns the intellectual property when new knowledge is created? How are copyrights, database rights, patent law, ethics, personal and human rights to be considered? It may be possible to apply a general public license (GPL)[iii] or other freeware license to research products created on a grid. A new licensing framework may be necessary for commercial products. Either way, this important issue needs to be resolved over the long term.

6. CONCLUSION

This paper discussed our experience with and perspectives on biomedical knowledge creation on the Open Bioinformatics Grid (OBIGrid) from the viewpoint of community development, service interoperability and intellectual property development as well as web services and workflow made available through grid computing.

Grid computing has great potential to become a platform for biomedical knowledge creation. The key to this knowledge creation is the transfer of knowledge from tacit knowledge to explicit knowledge. Ontology and mathematical modeling play essential roles in the representation of descriptive biomedical knowledge.

The knowledge spiral also requires "ba," a place in which people share knowledge and work together. Web services and workflows for bioinformatics help extend the community for the purposes of knowledge sharing. However, much remains to be done in terms of enhancing the interoperability of the services and protecting intellectual property rights that arise from the development of new knowledge.

[iii] http://www.gnu.org/licenses/

ACKNOWLEDGEMENT

The author expresses special thanks to Dr. Sumi Yoshikawa, Dr. Ryuzo Azuma and Dr. Kazumi Matsumura of RIKEN GSC for intensive discussions on drug interaction ontology and stochastic particle simulation. He also thanks Dr. Fumikazu Konishi, Mr. Ryo Umetsu and Mr. Shingo Ohki of RIKEN GSC and his students at Tokyo Institute of Technology for fruitful discussions and the implementation of the grid and web services on OBIGrid.

REFERENCES

[1] Collins F. S., Morgan M., and Patrinos A., The Human Genome Project: Lessons from Large-Scale Biology, Science, p.286, (11 April 2003).

[2] Nonaka I., Toyama R., and Konno N., SECI, Ba and leadership: a unified model of dynamic knowledge creation, Long Range Planning, Vol. 33, pp. 5-34, (2000).

[3] Kitaro Nishida. An Inquiry into the Good, translated by Masao Abe and C Ives. New Haven, USA: Yale University Press, (1990/1911).

[4] Konagaya Akihiko and Satou Kenji (Eds), Grid Computing in Life Science. Lecture Notes in Bioinformatics, Vol. 3370, (2005).

[5] Forbes GM, Micronutrient status in patients receiving home parenteral nutrition, Nutrition, Vol. 13, pp. 941-944, (1977).

[6] Melamed I., Bujanover Y., Igra Y. S., Schwartz D., Zakuth V., and Spirer Z., Campylobacter enteritis in normal and immunodeficient children, Am. J. Dis. Child, No. 137, pp. 752-753, (1983).

[7] The Gene Ontology Consortium, Gene Ontology: tool for the unification of biology, Nature Genetics, Vol. 25, pp. 25-29, (2000).

[8] Nagashima T., Silva D.G., Petrovsky N., Socha L.A., Suzuki H., Saito R., Kasukawa T., Kurochkin I.V., Konagaya A., and Schoenbach C., Inferring higher functional information for RIKEN mouse full-length cDNA clones with FACTS, Genome Research Vol. 13, pp. 1520-1533, (2003).

[9] Martin-Sanchez F., et al., Synergy between medical informatics and bioinformatics: facilitating genomic medicine for future health care, J. Of Biomedical Informatics, Vol. 37, pp. 30-42, (2004).

[10] Hatakeyama M., Kimura S., Naka T., Kawasaki T., Yumoto N., Ichikawa M., Kim J.H., Saito K., Saeki M., Shirouzu M., Yokoyama S., and Konagaya A., A computational model on the modulation of mitogen-activated protein kinase (MAPK) and Akt pathways in heregulin-induced ErbB signaling, Biochemical Journal, No. 373, pp. 451-463, (2003).

[11] Keuzenkamp-Jansen C.W., DeAbreu R.A., Bokkerink J.P., Lambooy M.A., and Trijbels JM., Metabolism of intravenously administered high-dose 6-mercaptopurine with and without allopurinol treatment in patients with non-Hodgkin lymphoma, J Pediatr Hematol Oncol., Vol. 18, No. 2, pp. 145-150, (1996).

[12] Ingelman-Sundberg M., The human genome project and novel aspects of cytochrome P450 research., Toxicol Appl Pharmacol., (29 June 2005).

[13] Yoshikawa S., Satou K., and Konagaya A., Drug Interaction Ontology (DIO) for Inferences of Possible Drug-drug Interactions. In: MEDINFO 2004, M. Fieschi et al. (Eds), IOS Press, pp. 454-458, (2004).

[14] Azuma R., Yamaguchi Y., Kitagawa T., Yamamoto T., and Konagaya A., Mesoscopic simulation method for spatio-temporal dynamics under molecular interactions, HGM2005 (Kyoto, Japan), (2005).

[15] Kecheng L., Incorporating Human Aspects into Grid Computing for Collaborative Work, Keynote at ACM International Workshop on Grid Computing and e-Science (San Francisco), (21 June 2003).

[16] Konagaya A., Konishi F., Hatakeyama M., and Satou K., The Superstructure toward Open Bioinformatics Grid, New Generation Computing, No. 22, pp. 167-176, (2004).

[17] Satou K., Nakashima Y., Tsuji J., Defago X., and Konagaya A., An Integrated System for Distributed Bioinformatics Environment on Grids., Springer, LNBI, Vol. 3370, pp. 8-18, (2005).

[18] Fukuzaki A., Nagashima T., Ide K., Konishi F., Hatakeyama M., Yokoyama S., Kuramitsu S., Konagaya A., Genome-wide functional annotation environment for Thermus thermophilus in OBIGrid, LNBI, Springer, Vol. 3370, pp. 32-42, (2005).

[19] Delcher A.L., Harmon D., Kasif S., White O., and Salzberg S.L., Improved microbial gene identification with GLIMMER Nucleic Acids Res., Vol. 27, No.23, pp. 4636-4641, (1999).

[20] Altschul S.F., Madden T.L., Schaffer A.A., Zhang J., Zhang Z., Miller W., and Lipman D.J., Gapped BLAST and PSI-BLAST: a new generation of protein database search programs, Nucleic Acids Research., Vol. 25, pp. 3389-3402, (1997).

[21] Chenna R., Sugawara H., Koike T., Lopez R., Gibson T.J., Higgins D.G., and Thompson, J.D., Multiple sequence alignment with the Clustal series of programs. Nucleic Acids Research, Vol. 31, pp. 3497–3500, (2003).

[22] Umetsu R., Ohki S., Fukuzaki A., Shinbara D., Kitagawa T., Hoshino T., and Konagaya A., An Architectural Design of Open Genome Services (OGS), Life Science Grid 2005 (Biopolis, Singapore), (2005).

CYBERINFRASTRUCTURE FOR THE BIOLOGICAL SCIENCES (CIBIO)

JOHN C. WOOLEY

Department of Pharmacology and of Chemistry-Biochemistry, University of California, San Diego,

9500 Gilman Dr., La Jolla, CA 92093-0446, California, USA

Email: jwooley@ucsd.edu

21st Century Biology is a product of dramatic advances in the life sciences as well as in many other disciplines, especially the computer and information sciences. To build and to prepare for the future of biology, one has to build a strong and stable infrastructure for the life sciences, termed the Cyberinfrastructure for the Biological Sciences (CIBIO). This all-pervasive CIBIO will go well beyond a network access, storage needs, grid computing infrastructure, databases and knowledge bases. It will pose great challenges in the way the life sciences are funded, how biology is taught in schools and universities, and how future biologists are trained. A complete change in the way we think about the life sciences is needed for a successful transformation of the old biology into the 21st Century Biology.

1. INTRODUCTION: THE CYBER CONTEXT

The USA National Science Foundation (NSF) recently introduced the term "cyberinfrastructure" (CI) to describe the integrated, ubiquitous, and increasingly pervasive application of high performance computing and advanced information technology (IT) approaches, which are already changing both science and society [1,2]. For academic research, storage needs complement network access; similarly, the revolutionary advances summarized in the term scientific computing (SC) are central.

The NSF, consistent with its description of its role in the scientific enterprise, asserted that "people, and their ideas and tools" are at the heart of CI [1]. Such a CI would address (a) the provision of routine, remote access to instrumentation, computing cycles, data storage and other specialized resources; (b) the facilitation of novel collaborations that allow new, deep interdisciplinary and multidisciplinary research efforts among the most appropriate individuals at widely separated institutions to mature; (c) the powerful and ready access to major digital knowledge resources and other libraries of distilled information including the scientific literature; and (d) other features essential to contemporary research.

The entire worldwide community of scientists has embraced CI and has joined a dialogue, even if much of the discussion concerns components or specifics of a mature CI, rather than one that is comprehensive or mature. Consequently, the reception of the concept of CI as a maturing, philosophical and practical perspective — on the profound revolution provided through today's integration of continuing advances in SC and Information Technology (IT) - has been truly remarkable.

Once fully implemented, a robust and comprehensive CI will contribute simultaneously in many ways to both science and society, and will do so through many different vehicles. The vehicles with the highest impact will include those that establish tools for capturing, storing, and managing data, tools for organizing, finding and analyzing the data to obtain information, and tools for integrating disparate aspects of that information. Also included are those that serve to connect experimental and theoretical research in order to deliver a knowledge repository for further considerations (or what for sciences like biology has been termed a synthesis). More generally, a CI should enable the creation of far more robust, widely distributed research teams (for which in the 1980s Bill Wulf, then President, NSF, provided the name "collaboratories" [NRC 1989] [2]), provide a platform integration of information from multicomponent systems, and offer new training environments.

The implementation of CI in incremental fashion and tailored to each discipline's needs by the funding agencies around the world offers an especial opportunity - a perfect fit - for the biological sciences. In every study in every discipline the importance of an international effort has restated the original vision by the leadership of NSF: "no one agency and no one nation could ever be locally successful: almost by definition, a cyberinfrastructure can only be created on an international scale" [3]. In particular, for the interests of the Life Sciences Grid community, the NSF asserts that the utmost priority needs to be assigned to the response of the biological sciences community, which will have to be carried out on an international scale.

2. THE UNFOLDING OF 21ST CENTURY BIOLOGY

At the same time as the revolution in computing, there has been a revolution of equal magnitude and impact in the life sciences, where life science is meant to include all basic biological science and the whole of biomedical science. As a consequence of this revolution, the life sciences can be said to be at a critical junction in their research history. Especially at levels from molecules to organisms, over several decades, every field of life science research has benefited from the philosophy of a highly focused, carefully conceived approach termed "reductionist." Thus, each sub-discipline or life science domain has been a part of the unexpected and extraordinary victories of this approach to probing the mechanisms by which biology proceeds. This approach begins with the selection of simple systems that can be readily characterized, and the establishment of biological models to serve as abstractions for the ultimate study of humans.

Concomitant with the world's entry into a new Century, commonly called by those as diverse as leading physicists and famous politicians to be the Century of Biology, life science has entered a new era of synthesis of biological wisdom from its many shards of data. The synthetic process aims to inform experimental choices and to steer thoughtful considerations by biologists about the future of life science research, and in doing so on a scale ranging across life; namely, ranging from macromolecules to ecosystems. To inform the process and deliver this synthesis, biological scientists must collect, organize and comprehend unprecedented volumes of highly heterogeneous, hierarchical information, obtained by different means or modalities, assembled and organized with different standards, widely varying kinds (types) of data itself with individuality, historicity, contingency and temporal or dynamic character, over vast scales of time, space and organizational complexity [2,5].

3. THE HISTORICAL PATH

While the role of computing in the physical sciences and engineering is far better known, through the efforts of the life sciences community, the agencies charged with funding biology invested early, and in numerous ways ranging from ecology and the Long Term Ecological Research Sites (LTERs) [6] to structural biology and the Protein DataBank of biomolecular structures (PDB) [7], in the advancing IT world that is now leading to a comprehensive cyberinfrastructure. As computational biology and bioinformatics have become a more integral part of basic and applied life science research, support for what are among the key steps toward building a cyberinfrastructure for the life sciences has grown within the private foundations (such as NSF) and all of the funding agencies especially in the USA.

There is an extraordinary opportunity, at this point in time, to consolidate those activities and build a compelling, integrated program, one that links biology to all of the other sciences and is established seamlessly, reaching out to every nation in the world and to every sub-discipline of the life sciences. In particular, building a cyberinfrastructure for the biological sciences requires an interface to all of the quantitative sciences, not just to the fields of computer science and engineering. We have already seen examples, such as the Biomedical Information Research Network (BIRN) [8], where the biomedical sciences have recognized the importance of IT for their efforts, and that CI activities should reflect all disciplines stretching out finally to health care. Numerous additional examples are certain to follow; these early investments will catalyze revolutionary change, not mere incremental improvements, around the world.

The properties of a mature CI will certainly be well suited for the cottage industry comprising the biological scientists working around the world. This proverbial "lock and key fit" arises on one hand from the recent introduction of grid services along with methods for data integration and other features of modern information technology, and on the other, with the advent of a biological research approach that is focused on a systems level that is integrative, synthetic and predictive. Such an approach is what biological scientists have termed genome-enabled biology, genomes to life, or the new biology, and given more formality through the NSF nomenclature, which is used in this manuscript; namely, 21st Century Biology. The vantage point gained by looking at research issues in biology from a synthetic point of view, including the characterization of interacting processes, and the integration of informatics, simulation and experimental analysis, represents the central engine powering the entire discipline.

Not only does 21st Century Biology absolutely require a strong cyberinfrastructure, but also, more than any other scientific domain, biology, due to its inherent complexity and the core requirement for advanced IT, will drive the future cyberinfrastructure for all science. Any major biological research endeavour must engage fully in setting the course, in establishing an architectural plan describing the specific needs of the biosciences, in assembling the parts, and building a full blown, highly empowering cyberinfrastructure for the entire biological sciences community that it sponsors and serves.

As our understanding of living systems increases, the inherent complexity of biology has become very obvious, so apparent as to approach a daunting challenge. Indeed, biological sciences encompasses on the 24 orders of time from atomic motions to evolutionary events, more than ten orders of space from molecules to neurons in large mammals, six to nine orders in the populations ranging from molecules to organisms, and a hierarchical organizational dimension of enormous variety, which can not be readily quantified but is obviously as diverse as the other parameters. In addition, just as calculus has served as the language of the physical sciences, information technology (informatics) will become the language of the biological sciences. Although biological scientists have already typically managed data sets up to the limit set by each generation's computing parameters (cycles, storage, bandwidth), the singular nature of observations, the individuality of organisms, the typical lack of simplifying symmetries and the lack of redundancy in time and space, the depth of detail and of intrinsic features distinguish biological data, rather than sheer volume.

The biological sciences, in settings around the world, will remain dominated by widely distributed, individual or small team research efforts, rather than moving to a particular focus on centralized, community facilities, as has happened for some sciences; the consequences of reaching out to the broadest range of the best performers wherever they are is, consequently, particularly important. As telecommunication networks advance, biologists around the entire world will be able to explore and contribute to 21st Century Biology.

At the molecular level, for example, a cyberinfrastructure for biology, using tools developed to extract implicit genome information, will allow biologists to understand how genes are regulated; how DNA sequences dictate protein structure and function; how genetic networks function in cellular development, differentiation, health and disease. In forming a CIBIO [4], the cyberinfrastructure ("CI") for biological sciences ("BIO") must integrate the expertise and approaches of scientific computing and information technology with experimental studies at all levels; for example, on molecular machines, gene regulatory circuits, metabolic pathways, signaling networks, microbial ecology and microbial cell projects, population biology, phylogenies, and ecosystems.

As the consequence of the parallel, fully comparable revolutions in biological research, and in computer and information science and engineering, an extraordinary frontier is emerging at the interface between the fields. Both communities, and for instance, their federal counterparts in the National Science Foundation, Directorate for Biological Science (BIO) and Directorate for Computer & Information Science (CISE), can facilitate the research agenda of each other. 21st Century Biology absolutely requires all of the insight, expertise, methodology and technology of advanced information technology (IT), arising from the output of computer science and engineering (CSE) and its interconnection with experimental research, or in other words, arising from the domain known as scientific computing (SC).

Indeed, only the biological sciences, over the past several decades, have seen as remarkable, sustained revolutionary increases in knowledge, understanding, applicability, as the computer and information sciences. Today, the exponential increases in these two domains make them ideal partners and the dynamics of the twin revolutions underpin the potential unprecedented impact of building a cyberinfrastructure for the entire biological sciences.

Building on these successes, the essence, for Biological Sciences utilizing Cyberinfrastructure (CIBIO) in empowering 21st Century Biology, is to "Keep Your Eye on the Prize":

- Invest in People to Nurture the Future
- Ensure Science Pull, Technology Push
- Stay the Course
- Prepare for the Data Deluge
- Enable Science Targets of Opportunity
- Direct Technology Contributions
- Establish International Involvement

In other words, the BIO component of the CIBIO must provide the vision for the CIBIO, and not rely on technology drivers. In the case of the NSF, once involved, its BIO directorate will have made a major commitment to the community and must have an effective long-range plan to sustain the efforts. The changing relationship of an agency funding computer and information science, such as that at the NSF CISE, to its high performance computing centers and the introduction of a CI process across the board, places a significant obligation on biological sciences research funding agencies such as the NSF Biological Sciences Directorate to structure and maintain the role of the biological science community in the development and utilization of scientific computing and information technology applied to biology.

The most obvious feature of 21st Century Biology is the increasing rate of data flow and simultaneously, the highly complex nature of the data, whether obtained through conventional or automated means. This immediately highlights the value of approaching the funding of 21st Century Biology on the premises of CIBIO.

Not all sub-disciplines can be simultaneously provided with a CI by BIO, so selected pilot projects and areas of high biological impact should be the first focal points of effort. Nothing succeeds like success, and the complete implementation of a CI for the biological sciences will depend on the initial choices paying off in easily demonstrated ways. Thus, the early pilots should also be selected for their ability to contribute significantly in the near term, even though many aspects of a comprehensive CI for the biological sciences will take years to develop fully.

While all communities should interoperate as a whole to seek to absorb as many as possible of the computational contributions from other fields rather than encouraging reinvention, BIO must also choose its own technology course, not passively accept whatever (hardware, software, middleware) is delivered for the needs of other science domains.

Scientists can now facilitate the progress of each other in extraordinary ways. To optimize introduction of 21st Century Biology, the biosciences need to be interconnected to the other scientific domains as well. So in the long run, if we underestimate the importance for biology or fail to provide fuel over the journey, it would be damaging, perhaps even catastrophic, for the entire community.

4. SCIENCE AND TECHNOLOGY SERVING SOCIETY

Cyberinfrastructure promises to be as pervasive and central an influence as any societal revolution ever. Given the breadth and the long-term impact, several considerations are very important. First, working with partnerships and working in a global context is obvious and imperative on a scientific basis. Second, these interconnections are equally obvious and imperative on a practical and administrative basis. The cost of full implementation, of a comprehensive cyberinfrastructure in which the biological sciences benefit from cyber-rich environments, such as those piloted by the Network for Earthquake Engineering Simulation (NEES) [9] and BIRN [8], will be large as would be expected for incredible significance and applicability. This will be a decades-long effort. Nonetheless, even every scientific journey begins with a single step and the scientific community should just take that step (and further initial steps no matter how small each might have to be), as soon as possible.

Funding increases for the frontier at the interface between computing and biology will obviously be needed as well to extend the experimental science projects to permit them to fully exploit the cyberinfrastructure and to build collaborations for the synthetic understanding of biology, which requires computational expertise and deep involvement of information technology. Only those nations that recognize that the economic and health care implications of the life sciences justify the growth in annual expenditure will remain competitive; indeed, any nation that fail to grasp and implement CI will miss the societal contributions from entire scientific revolution at the interface.

Beyond this base, major partnerships with computer and information science and with the other sciences will be required. The impact should not be underestimated but neither should the requirement for greatly enhanced, stable funding. The implementation of some preliminary features of a CIBIO followed by termination of the effort would have a serious effect on productivity in the life sciences.

Already, the NSF BIO directorate is already engaged upon a series of extraordinary opportunities, in creating a larger scale for shared, collaborative research efforts, through activities like Frontiers in Integrative Biological Research (FIBR) [10] and National Ecological Observatory Network (NEON) [11], while sustaining microbial projects and LTER. These larger scale projects particularly require a cyberinfrastructure, with costs of comparable magnitude to the projections for experimental research. Such major leading initiatives in USA will have direct impact and consequence on the biological and biotechnology initiatives in countries worldwide.

The Life Sciences funding programs at the NSF and around the world will have to (a) build up their own core activities at this interface (e.g., the funding for bioinformatics, biological knowledge resources, computational biology tools and collaborations on simulation/modeling) that allow it to partner with other parts and other branches of their scientific funding activities, (b) choose test beds for full implementation of CI, establish paths toward deep integration of CI into all of its communities and for all of its performers, and (c) set a leadership role for other agencies around the entire world to follow. Only through a decades long commitment and through flexible, agile, engaged, proactive interactions with the entire community and with the other stakeholders – i.e., with other sources of funding for science – will the effort be a complete success.

Two categories of early actions are needed. The first implementation steps should be to expand the extant database activities and computational modeling/simulation studies, which already need far more focused attention than they have gotten in the past decade. Many obstacles remain for individual grand challenge research problems as well as for the particularly difficult implementation problems for the establishment of an ever larger number of databases in the life sciences and their subsequent distributed integration on demand to underpin experimental and theoretical work in the field. The provision of increased funding for algorithm development and for teams coupling experimental probes with modeling would allow simulation studies to contribute considerably more across all life sciences. Accelerating the introduction and expansion of tools and of the conceptual approaches provided through testing models, a prominent feature of research in the physical sciences, will require continued programmatic emphasis and commitment.

This renewed focus is essential for 21st Century Biology, in that many biologists trained in more traditional ways are just starting to recognize the opportunities. Encouragement of more collaboration between/among experimentalists and computational scientists is essential, but the full implementation of the opportunities will require the training of a new generation of translators, of "fearless" biologists able to understand and speak the language of the quantitative scientists well enough to choose the best collaborators and to build bridges to more traditionally trained experimentalists. Many basic requirements involve academic professionals and the use of well-documented approaches within computer and information science. Implementing these important requirements will be the responsibility of biological funding agencies and must be in place for effective collaborations on research frontiers with the other scientific funding agencies.

Other early actions are to establish a long range plan for sustained funding and to engage the community in a dialogue to ascertain implementation priorities as well as to prepare the biological scientists from around the world to participate fully. The enabling and transformational impact of CI justifies and for full implementation would require an increase in funding, but it will also require that biological funding agencies to lead a much larger effort, marshalling resources from other agencies around the world, to provide adequate funding to ensure full participation by the international life sciences community.

Important administrative features include the review and funding of infrastructure and establishing (over time) a balance across the sub-disciplines. Infrastructure is different from individual research and needs separate processes for their consideration. Central coordination, needed for effective selection of pilots and coordinated efforts, will ensure balance and accelerate penetration of the benefits of modern IT to every bio-scientific discipline.

All categories of infrastructure are increasingly important for scientific research, but cyberinfrastructure will be particularly valuable for the biological sciences. What will be critical is to recognize that infrastructure cannot be treated the same as individual research proposals. One cannot review infrastructure against individual research, and separate, centralized review and oversight will be needed. Infrastructure benefits all, but has a different time frame, different budgets, different staffing (more academic professionals), and can not be simultaneously considered with individual projects. At the same time, robust, rigorous peer review is essential to establish the best opportunities. Competition is also important; overlapping efforts will need to be initiated in many cases and then the best project will ultimately become clearly identified.

The educational challenges are themselves vast, and will require an expansion of existing programs and possibly the creation of new ones. CI will be dramatically alter how education is conducted - the means for training and transferring knowledge – and its full implementation and utilization will require a new cadre of scientists adroit at the frontier between computing and biology, able to recognize important biological problems, understand what computational tools are required, and capable of being a translator or communicator between more traditionally-trained biologists and their collaborators, computational scientists who will be just as traditionally-trained. These requirements are universal; that is, any bioscience funding should work with generic scientific agencies and with international agencies to encourage innovation and sustain the excitement beyond disciplinary and national boundaries.

Within a single country, we can also consider the distribution of effort and synergistic cooperation. Take the example of the USA. Its National Institutes of Health (NIH) will inevitably need to take responsibility for the CI for biomedical, translational and clinical medicine and health care. The USA Department of Energy (DOE) will need to build a CI to connect its laboratories and to promote energy and environmental applications of research in biology. With such cooperation, whilst the USA NSF may be able to nucleate an activity, it may not have to plan for long term, expanding support as these other agencies may even sustain some or much of the original core.

Some research problems, such as ecology, plant science, phylogeny and the tree of life, the evolution of multicellularity, and of developmental processes, among others, are areas that biological sciences will always own. Besides CI applied to these categories of very basic biological science research, for the foreseeable future, the overall catalysis of life science by CI will remain a role for those agencies, around the world, responsible for funding basic and applied biology. They must ensure that once the cyberinfrastructure for biology is put in place, funds must be in place to sustain the efforts and in particular, budget presentations must be in place to ensure that prototype and pilot efforts, after selection of the best activities, can be funded and maintained stably to deliver for the community.

REFERENCES

[1] Atkins, D. (Chair) "Revolutionizing Science and Engineering through Cyberinfrastructure; Report of the Blue-Ribbon Advisory Panel on Cyberinfrastructure" National Science Foundation, Arlington VA. (January 2003) http://www.nsf.gov/od/oci/reports/toc.jsp

[2] Cerf, VG et al., National Collaboratories: Applying Information Technologies for Scientific Research, National Academy Press, Washington, D.C. (1993) http://www7.nationalacademies.org/cstb/pub_collaboratories.html

[3] NSF's Cyberinfrastructure Vision for 21st Century Discovery. NSF Cyberinfrastructure Council, National Science Foundation. CI DRAFT: Version 7.1 (July 20, 2006) http://www.nsf.gov/od/oci/ci-v7.pdf

[4] Wooley, JC and Lin, HS. (Eds) Catalyzing Inquiry at the Interface between Biology and Computing. Committee on Frontiers at the Interface of Computing and Biology, Computer Science and Telecommunications Board, Division on Engineering and Physical Sciences, National Research Council of the National Academies, USA. National Academies Press, Washington DC. (2005) http://www7.nationalacademies.org/CSTB/pub_biocomp.html

[5] Subramaniam S and Wooley, J. DOE-NSF-NIH 1998 Workshop on Next-Generation Biology: The Role of Next Generation Computing. (1998) http://cbcg.lbl.gov/ssi-csb/nextGenBioWS.html

[6] Callahan, JT. Long-Term Ecological Research. BioScience, **34**(6): 363-367 (1984)

[7] Berman, HM, Westbrook, J, Feng, Z, Gilliland, G, Bhat, TN, Weissig, H, Shindyalov, IN, Bourne, PE. The Protein Data Bank. Nucleic Acids Research, **28**:235-242 (2000).

[8] http://www.nbirn.net/

[9] http://www.nees.org/

[10] http://www.nsf.gov/pubs/2003/nsf03581/nsf03581.htm

[11] http://www.neoninc.org/

UPCOMING STANDARDS FOR DATA ANALYSIS IN BIOINFORMATICS

MARTIN SENGER, TOM OINN, PETER RICE

EMBL Outstation – Hinxton, European Bioinformatics Institute, Wellcome Trust Genome Campus, Cambridge CB10 1SD, UK

The life sciences research domain depends on effective access to the data and data analysis. There are tool packages (e.g. EMBOSS) and service repositories (e.g. BioMoby) that bring together a critical mass of computational methods. A standard defining how these tools can be integrated into data analysis grids is needed. The paper describes an upcoming standard for Life Sciences Analysis Engine that facilitates uniform access to data analysis operations. The paper also points to the early implementation of the standard: Soaplab and its companion Gowlab.

1. INTRODUCTION

In the Life Sciences domains, there are many computer methods for analyzing and deriving data. These methods constitute a vast "knowledge grid" with well defined "nodes" but often only vaguely defined interconnecting "edges". Because the ideal of a uniform, ultimately connected world in science is unrealistic, the main challenge is to help scientists to discover those resource nodes and to integrate them to form aggregate experimental functionality despite their heterogeneity.

The integration does not come for free but the new tools grow with the awareness that their strength is not only in their scientifically based algorithms but also in their ability to integrate and to be integrated with other tools for data analysis. A splendid example of this trend is EMBOSS [1], a "free Open Source software analysis package specially developed for the needs of the molecular biology user community". The way in which the software intelligently copes with data in a variety of formats and how it transparently allows the retrieval of sequence data from the web makes it a standard in itself.

From the perspective of the grid computing and data analysis integration, the EMBOSS is not only a scientifically valuable package but can also bring about a 'critical mass' at which point tool integration becomes effective. It comprises more than 140 individual analysis tools, and additionally allows the same or similar access to over 100 additional programs from third parties, wrapped as EMBASSY tools.

Another resource that can provide this critical mass of available tools, in this case exposed as Web Services, is BioMoby [2,3], a system allowing interaction with multiple sources of biological data regardless of the underlying format or schema. BioMoby also "allows for the dynamic identification of new relationships between data from different sources" – it provides a discovery framework whereby various tools may be added in a way which guarantees that they will be inter-operable with other BioMoby tools. The central BioMoby registry has, at the moment of writing, more than 120 interoperable services.

The last resource, but not the least, is, of course, the Internet itself. There is a virtually unlimited set of available scientific data analysis tools exposed through interactive web interfaces. While not readily amenable to systematic integration the sheer quantity, variety and richness of scientific data they represent is intrinsically attractive.

These three resources were, from the very beginning, the prime focus for integration efforts carried on within the myGrid [4] project; this project focuses on finding, adapting or developing all fragments crucial for effective integration within an e-science context. While the myGrid project encompasses a variety of integrative technologies such as distributed query processing, semantic service discovery and support for virtual organizations, the primary focus of this paper is the mechanism by which domain specific analysis and data services as described above may be accessed programmatically; this uniform access mechanism in turn allows the integration of these tools into more complex functional assemblies. This paper also represents an active contribution to various standardization efforts, namely the W3C, OMG and I3C consortia.

2. LIFE SCIENCES ANALYSIS ENGINE

The effort to establish a standard for controlling analysis tools within a distributed environment has been active for some time. Regarding standardization bodies involved in the life sciences domains, in January 2000 the OMG (Object Management Group[1]) adopted the Biomolecular Sequence Analysis specification [5] containing the first standard for a general analysis engine. Although labelled

[1] The Object Management Group (OMG) is the world's largest software consortium with an international membership of vendors, developers, and end users. Established in 1989, its mission is to help computer users solve enterprise integration problems by supplying open, vendor-neutral portability, interoperability and reusability specifications based on Model Driven Architecture (MDA).

"biomolecular sequence analysis", the specification was quite general. This standard was defined for CORBA, a favourite architecture of that time and one of its successful implementations was AppLab [6].

Time went on, and the OMG started promoted MDA, the Model Driven Architecture [7]. MDA defines an approach to IT system specification that separates the specification of system functionality from the specification of the implementation of that functionality on a specific technology platform, and provides a set of guidelines for structuring specifications expressed as models. The motto is "model it once, use it with as much middleware as you need". The MDA expresses a domain knowledge in a PIM, a Platform[2] Independent Model, and technology dependent standards in PSMs, the Platform Specific Models (an example would be CORBA).

In 2004, the OMG technology adoption process is finalizing a new, MDA-based standard for the Life Sciences Analysis Engine.

The goal of this specification is to describe how to access data analysis tools in an inter-operable way within a distributed environment. It includes the execution and control of computer methods (represented as applications, programs, processes, threads, or otherwise), and dealing with their inputs, output, and functionality. Particularly, it addresses:

- Both synchronous and asynchronous invocation of the remote tools. This reflects the common need to run time-consuming bioinformatics tools that can last for hours in a completely automatic way. It is a fundamental requirement for efficient and robust tool integration.
- The ability to transport data to and from the remote tools both in its entirety and in smaller chunks. Again, it comes from the real use cases of analyzing large genomic data sets.
- The ability to interact with the tools that have already been started. It allows for interactive simulation of biological processes – it does not fulfil all requirements usually expected from a real-time system, but it comes close.

The specification also stresses the importance of producing verifiable data by introducing interfaces and data structures that provide provenance information pertaining to the analysis. Provenance is understood as the information about the origin, version, and way of usage of an analysis tool, or of the data used by the tool, and the process of tracing and recording such information.

---

[2] According to the OMG definition, "platform" is a set of subsystems/technologies that provide a coherent set of functionality through interfaces and specified usage patterns that any subsystem that depends on the platform can use without concern for the details of how the functionality provided by the platform is implemented.

3. DOMAIN MODEL FOR AN ANALYSIS ENGINE

The interfaces and data structures describing the domain model are separated into several parts:

- Metadata describing analysis tools and interfaces to access them
- Interfaces and structures to control analysis tools
- Interfaces and structures for notification events
- Provenance data description

Metadata used to describe analysis tools can be evil! The eternal problem with metadata is that they are expected to be flexible and extensible, and at the same time to be highly interoperable. Any solution for this dilemma must also mandate the location in which these metadata are defined. One approach is to define metadata within the API of the analysis services; such a solution makes metadata more explicit, more interoperable, but also much less flexible. Alternatively they can be defined independently from the analysis interfaces, for example as an XML DTD or XLM Schema, supposedly shared by all platform-specific models; these solutions allow metadata to be very flexible and extensible because they are not part of the MDA architecture at all, but this also makes them less interoperable.

With an awareness of the metadata dilemma sketched above, the analysis engine specification treats metadata in the following schizophrenic way:

- It defines a platform-independent, minimal metadata model that includes the most expected metadata. It contains metadata that are not only useful for the service providers but for the clients (for example, metadata describing available parameters of an analysis service can be used to build a GUI for such service).
- It does not, however, propagate this minimal metadata model to the specific platforms. The metadata are expressed there as a simple data type (such as an XML string), expecting that this data type contains at least what is defined in the minimal metadata model without specifying exactly how it is contained. The implementation must document this containment; we believe that it would be unrealistic to expect more.

The minimal metadata model details are in the specification; here we present only a figure sketching the major entities. The core entity describes "parameters". The name "parameter" is a legacy artefact referring to the command-line tools many analyses operations are based on, but its semantics are quite generic. Each parameter describes a piece of data that is either entering a given analysis tool, or that is produced by the tool. Typically a parameter describes:

- The type and format of the entering data. It can describe a small piece of data, such as a command-line option, or a real data input, such as an input file.
- How to present this piece of data to the underlying analysis tool. For example, a command-line tool can get input data on its command line, or it can read them from its standard input.
- Constraints for the values of these data and how they depend on values of the other parameters.
- The type and format of the data produced as output, and how they can be obtained from the underlying analysis tool (e.g. in a file, or by reading analysis' standard or standard error output).

The life cycle of a single invocation of an analysis tool consists of the creation of a job, feeding it with input data, executing it, waiting for its completion (or interrupting it earlier) and fetching resulting data. It can be accompanied by event handling based on observation of changes in the execution status of the job.

Each analysis tool is represented by an Analysis_service object. When a service is executed (using a set of given input data) it creates a Job object that represents a single execution of the underlying analysis tool. Each Analysis_service object can create many jobs during its lifetime, each of them with a new set of input data.

The main goal of the Analysis_service and Job interfaces is to allow:

- Fetching all information about an analysis tool
- Feeding the analysis tool with input data in an appropriate format
- Executing the tool either in a synchronous (blocking) way or in an asynchronous way
- Receiving notifications from the running tool about its status
- Fetching resulting data produced by the tool at an appropriate time

Any analysis client may express an interest in state changes within a running job. A prominent change in the job life-cycle is, of course, when the job terminates (either by an error or by the user request or by successful completion of its task). The interfaces for controlling an analysis tool provide two methods by which information about job changes may be conveyed to the client:

- client-poll notification
- server-push notification, represented by a notification channel

The client-poll is mandatory for compliance with the specification, the server-push not because for some platform-specific implementations it may be difficult to achieve it.

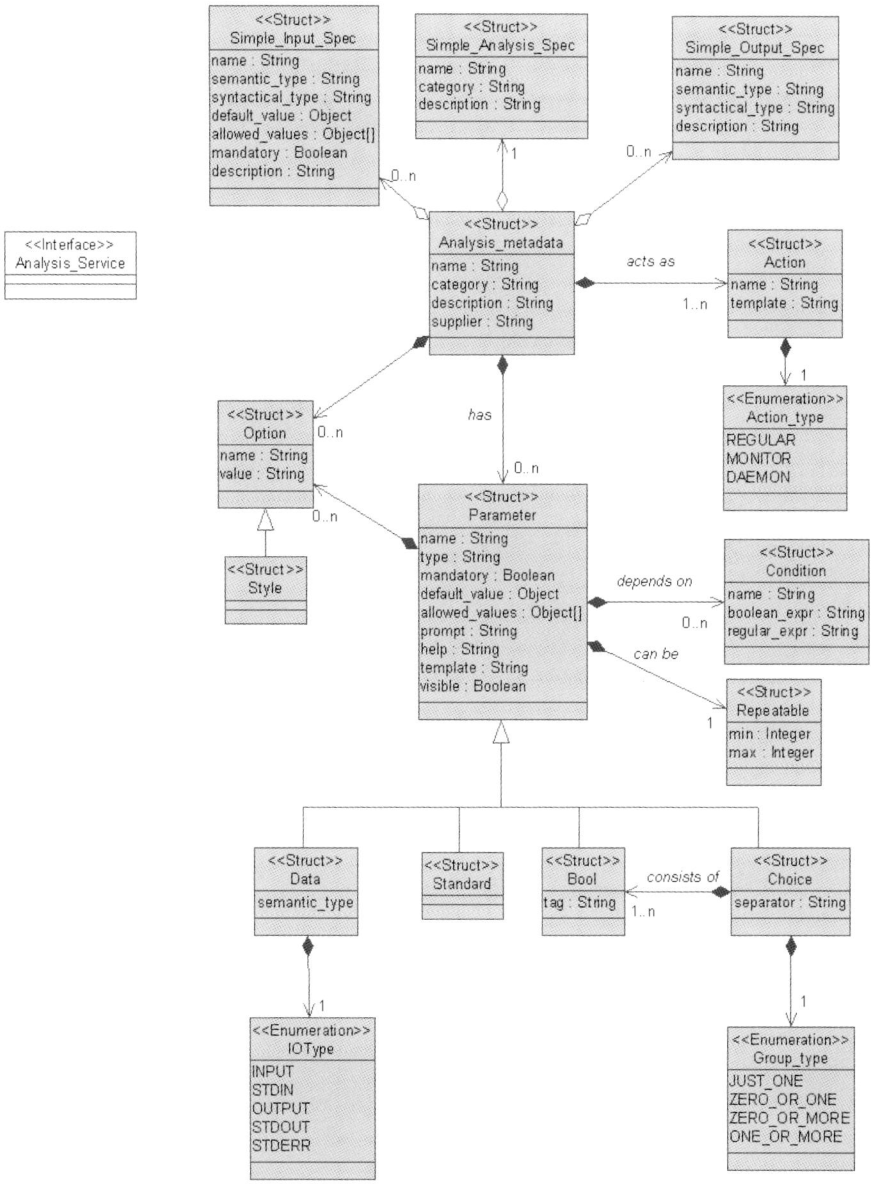

Figure 1. Minimal metadata model.

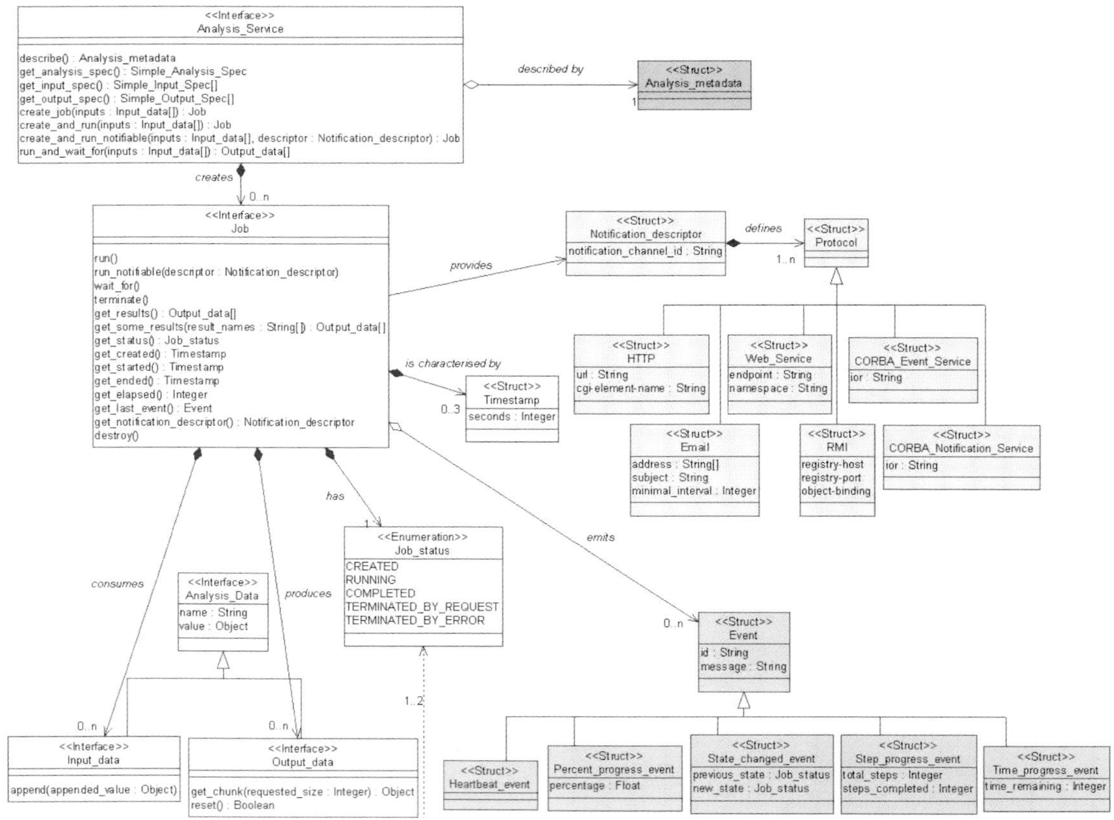

Figure 2. Analysis Service model.

Although these mechanisms differ in the exact way the events are being exchanged, both share the same format for the events themselves. The notification channel mechanism is the more complex (and more powerful) way: it uses another service, or even set of services, to transport event messages (which may involve things like setting an expiration time, secure channels, or postponed and re-tried deliveries etc.). The specification allows the use of existing standards for the notification, or use of home-made ones. The mechanism used depends on the result of the client-server negotiation.

4. TECHNOLOGY SPECIFIC MODELS

The MDA approach allows creation of one or more middleware-specific models derived from the platform independent model. Although this does not always guarantee interoperability between various implementations it goes close to it. The Life Sciences Analysis Engine defines interfaces for Java™ (letting the implementation define the specific network protocol to use), and, perhaps more interesting for grid computing, a set of Web Service interfaces with a binding for SOAP over HTTP.

The principles applied during the conversion from PIM to PSM guarantee that the resulting data types are suitable for the Web Services architecture in many programming languages. These principles are as follows:

- Used data types are simple in order to avoid as much as possible the need for implementing specific data mappings.
- The state handlers are part of the method signatures in order to be less dependent on the session management in any particular programming language and/or Web Services toolkit.

5. EARLY IMPLEMENTATION

There is no proof that any standard is useful until it is implemented and used. The myGrid participants were the early developers of an implementation of the new Life Sciences Analysis Engine – a project called Soaplab [8]. Simply speaking, Soaplab is a set of Web Services providing a programmatic access to analysis tools on remote computers. It implements the Web Services based platform specific model of the Life Sciences Analysis Engine.

Soaplab itself does not integrate – but by applying standards methods it allows integration. It facilitates access to all EMBOSS and other programs in a unified way.

Accessing command-line analysis tools is powerful but not good enough. As stated in the introduction, the richest pool of information resources is available as ordinary web pages, static or dynamic. They are mostly designed to be accessed by humans and their clicking fingers; however, their abundance is tempting. Therefore, Soaplab (using its subproject called Gowlab) applies the same interfaces as described earlier as a front-end to the web resources. Of course, there is no free lunch. The web pages are very non-standardized and they tend to change often. It places higher demands on the providers running Gowlab-based services, but the Gowlab system tries to help as much as possible with creating and plugging-in new HTML parsers. The reward is that the end users get suddenly - without changing anything in their client programs - vast amount of new resources.

Although outside the scope of this paper, it is worth mentioning the Taverna [9] project – a subproject of myGrid focused on the construction, manipulation and enactment of workflows. It is clearly the case that without a critical mass of services, provided through this specification in the form of the Soaplab server running on top of the EMBOSS tool set, this tool would not have been adopted as rapidly as it has. All integration platforms carry some overhead for their end users in terms of learning curve and existing work which requires re-factoring to fit within the new platform – the presence of some hundreds of analyses via Soaplab is a compelling incentive to make this investment of time and effort. In turn, the existence of functioning workflow platforms such as Taverna is a compelling argument for the provision of services through this specification – ‘jump started’ by the mass of applications within the EMBOSS package this mutually reciprocating cycle has resulted in a rapid growth in the number of available services to our end user community.

6. CONCLUSIONS

- The knowledge grid in bioinformatics consists of hundreds of data analysis tools.
- Their integration into data analysis grids and user-driven workflows is hard to achieve without standard access methods.
- Life Sciences Analysis Engine provides such standard. It allows several different approaches to the analysis tools (synchronous, asynchronous, interactive, etc.).
- The standard includes a general, technology-independent model that can be re-used by many different network technologies.

REFERENCES

1. Rice P., Longden I., Bleasby A., "EMBOSS: The European Molecular Biology Open Software Suite", Trends in Genetics, June 2000, Vol. 16, No 6. pp.276-277
2. Wilkinson M.D., Links M., "BioMoby: an open-source biological web services proposal", 2002, Briefings In Bioinformatics 3:4. pp 331-341.
3. Wilkinson M.D., Gessler D., Farmer A., Stein L., "The BioMoby Project Explores Open-Source, Simple, Extensible Protocols for Enabling Biological Database Interoperability", 2003, Proc Virt Conf Genom and Bioinf (3):16-26. (ISSN 1547-383X)
4. Stevens R., Robinson A., Goble C.A., "myGrid: Personalised Bioinformatics on the Information Grid" Proceedings of 11th International Conference on Intelligent Systems in Molecular Biology, 2003; Bioinformatics, Vol. 19 Suppl 1 2003, pp. 302-304
5. Object Management Group, "Biomolecular Sequence Analysis Specification, Version 1.0", http:// www.omg.org/cgi-bin/doc?formal/2001-06-08
6. Senger M., "AppLab – A CORBA-Java based Application Wrapper",1999, http://www.hgmp.mrc.ac.uk/CCP11/CCP11newsletters/CCP11NewsletterIssue8.pdf
7. Object Management Group, "MDA – Model Driven Architecture", http://www.omg.org/mda/
8. Senger M., Rice P., Oinn T., "Soaplab - a unified Sesame door to analysis tools", Proceedings of UK e-Science, All Hands Meeting 2003, Editors - Simon J Cox, pp. 509-513, (ISBN - 1-904425-11-9), 2003.
9. Oinn T., Addis M., Ferris J., Marvin D., Senger M., Greenwood M., Carver T., Glover, K. Pococl, M.R., Wipat A., Li P., "Taverna; A Tool for the Composition and Enactment of Bioinformatics Workflows", Bioinformatics Journal, 2004.

PARALLEL AND PIPELINED DATABASE TRANSFER IN A GRID ENVIRONMENT FOR BIOINFORMATICS

KENJI SATOU

School of Knowledge Science, Japan Advanced Institute of Science and Technology, 1-1 Asahidai, Nomi, Ishikawa 923-1292, Japan
Email: ken@jaist.ac.jp

SHIN'ICHI TSUJI

NEC Software Hokuriku, Ltd., 1 Anyoji, Hakusan, Ishikawa 920-2141, Japan
Email: s_tsuji@jaist.ac.jp

YASUHIKO NAKASHIMA

NEC Software Hokuriku, Ltd., 1 Anyoji, Hakusan, Ishikawa 920-2141, Japan
Email: hiko@jaist.ac.jp

AKIHIKO KONAGAYA

Riken Genomics Sciences Center, 1-7-22 Suehiro, Tsurumi, Yokohama, Kanagawa 230-0045, Japan
Email: konagaya@gsc.riken.jp

As a part of our Grid environment for bioinformatics (OBIEnv), we developed a database transfer tool *obiupdate* which facilitates easy, reliable, and efficient update of biological databases by a simple configuration. Among various features implemented in *obiupdate*, parallel and pipelined transfer is a key issue in an efficient transfer of database in various computers and network environments including broadband and narrowband WANs, 10/100/1000Mbps LANs, single node at a site, and multiple cluster nodes. From the experiments conducted on the real Internet environment, it was revealed that in comparison with naive transfer, *obiupdate* can efficiently spread a database to computers scattered at multiple sites by avoiding the problem so called flash crowds, which is a serious problem not only in popular Peer-to-Peer file sharing but also in biological database transfer.

1. INTRODUCTION

Today's biosciences are highly dependent on various databases on computers. Even in the case of most traditional biologists, daily use of web services like internet search and bibliographic search are essential. In addition, it is necessary for a bioinformatician to have a local copy of the latest version of a database to conduct an *in-silico* analysis with his/her original program. Such mirroring of biological databases has been done by computer centers related to biosciences. For example, the Bioinformatics Center at Kyoto University has been maintaining over 20 well-known databases which include databases from abroad (such as GenBank, SWISS-PROT, and PDB), databases from Japan (such as JSNP and PMD), and databases originally developed at the center (such as KEGG and aaindex). Such databases except original ones are first mirrored from various sites, then mirrored again in other computer centers and laboratories. Moreover, due to the remarkable improvement on the performance of personal computers on desktop, these databases are frequently copied by a significant minority of researchers.

Database mirroring is essential to bioinformatics. However, it is still not sophisticated and has the following problems:

- Unlike usual ftp mirror, a biological database mirroring requires transfer completion check, deployment of transferred database, and removal of old one. It should be performed automatically, but in typical centers and laboratories, it is manually done by administrators.
- Since the latest version of a database is first provided by one primary site (e.g. NCBI for GenBank), most people attempt to download it from the site without searching for the same copy on anywhere else. It causes a congestion of network traffic (so-called flash crowds, which is a serious problem also encountered in popular Peer-to-Peer file sharing).
- Different from mirroring a single file, it is needed to transfer multiple files with various sizes and formats (text and binary) for biological database mirroring. In case of a large database like GenBank, the number of entry files amount to over 600. In addition, if we need to mirror sequence files in FASTA format and index files for BLAST search, total number of files amounts to 4600 or more. In such a case, it is difficult to complete all transfers by only one execution of mirror program.
- It is well known that a single-threaded file transfer between two computers is really inefficient since it cannot fully utilize a network bandwidth. However, attempts to perform parallel transfer from primary site further escalate the flash crowd problem.
- Servers and clients of database mirroring are heterogeneous in their CPU power and network bandwidth. So, a method of transfer acceleration like compressed transfer is not always effective. In some cases, it makes a transfer slow by consuming much CPU power.

To solve these problems, we have developed a database transfer tool *obiupdate* as a part of Grid environment for bioinformatics OBIEnv reported in the previous study.[1] *obiupdate* facilitates easy, reliable, and efficient update of biological databases by a simple configuration. Among various features implemented in *obiupdate*, parallel and pipelined transfer is a key issue in an efficient transfer of database in various computers and network environments including broadband and narrowband WANs, 10/100/1000Mbps LANs, single node at a site, and multiple cluster nodes. From the experiments conducted on the real Internet environment, it was revealed that in comparison with naive transfer, *obiupdate* can efficiently spread a biological database to computers scattered at multiple sites by avoiding the flash crowds.

2. GRID ENVIRONMENT FOR BIOINFORMATICS

2.1. OBIGrid and OBIDuc

Open Bioinformatics Grid (OBIGrid) is one of the comprehensive Grid projects in Japan.[2,3] Aimed at the construction of a robust infrastructure for bioinformatics, OBIGrid has adopted a virtual private network (VPN) in order to guarantee transparency and security issues. Several research and development projects are presently running on it. As seen in Figure 1, VPN tunnels in OBIGrid are organized as

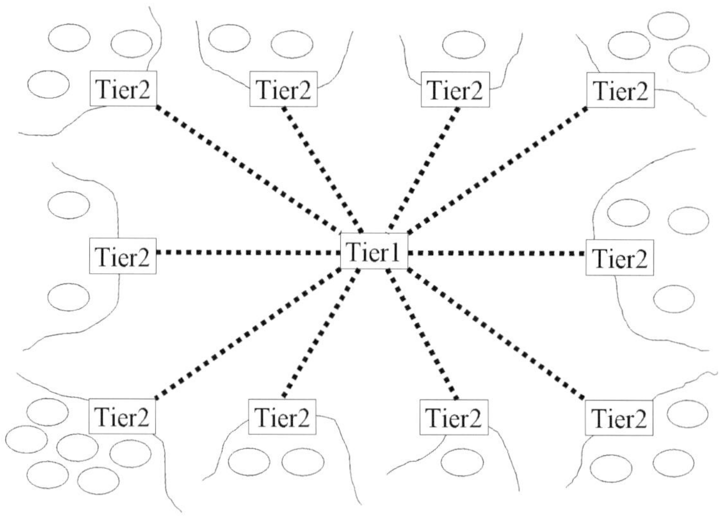

Figure 1. Basic architecture of OBIGrid. VPN routers (Tiers 1 and 2) and computers (nodes) are drawn as boxes and ovals, respectively. Dotted lines represent VPN tunnels.

Figure 2. Location of OBIDuc sites in Japan. The Tokyo area is magnified. Academic and commercial sites are pointed by light-gray and black pushpins, respectively. One site with medium-gray pushpin is a nonprofit organization (NPO).

star topology. It means that two nodes in different sites communicate with each other via VPN tunnels. Most of the Tier 2 routers are WatchGuard Firebox V10 with 20Mbps VPN throughput (3DES encryption). In addition, a few sites adopt V60, V80, and products by other vendors. On the contrary, Tier 1 router is a customized PC with single Xeon 2.8GHz CPU, 512MB memory, 35GB HDD, Intel PRO/1000 MT Dual Port Server Adapter, and NITROX XL NPB (CN1230-NPB), which is an encryption/decryption accelerator board produced by Cavium Networks. The Tier 1 router has about 900Mbps VPN throughput (3DES encryption).

In this study, we used OBIDuc[4] (Distributed but Uniform Cluster for Open Bioinformatics), which is an account-level overlay network on OBIGrid. Currently, 19 sites in Japan are joined to it (Figure 2), involving computers located at 18 working sites.

2.2. OBIEnv

OBIEnv[1,5,6] is an integrated system developed on OBIGrid. It facilitates distributed, high-throughput, adaptive, and transparent computing, as well as database management in bioinformatics. Moreover, robustness to various failures of node and network is well considered in its design and implementation.

In OBIEnv, an information server (P2P server) keeps track of information about each node. For instance, information about hardware, software, and database in a node is reported to the server. Based on the information, *obidispatch* divides and distributes a set of tasks given by a user with constraints on the execution of the tasks. During the parallel and transparent execution of tasks by *obidispatch*, the number of tasks currently dispatched to a node is periodically reported to the information server.

A simple configuration file *.obienv.conf* in the home directory of a special account name *obienv* is used to control the behavior of OBIEnv software at the node. For example, P2PSERVER (FQDN of information server), USERLIST (a list of account names allowed to use *obidispatch*), and DBLIST (a list of database names to be updated on the node) are specified in .obienv.conf. In addition, NETPATH (rough estimation of network path where the node is placed in a LAN) is included in it. For instance, if a path from a node to WAN is estimated as "node - switch A - switch B - router ", the value of NETPATH in .obienv.conf can be specified as "switchA.switchB.router". More intuitively, it is sufficient to represent the levels of bandwidths in a LAN. If a LAN has the same bandwidth at everywhere in it (or, at least a user has estimated so), the simplest description with only one level is sufficient even if the LAN consists of a hierarchy of cascaded network components.

2.3. *Obiupdate*

obiupdate is a database transfer tool included in OBIEnv. Aside from parallel and pipelined transfer described in the next section, it has the following functions:

- Simple configuration by DBLIST in .obienv.conf.
- The latest version of the specified database is automatically searched, transferred, and deployed.
- Multiple versions of a database can be maintained (by default, the number of versions for a database to be kept in a node is set to 1).
- Double invocation of *obiupdate* is avoided. This function is needed to maintain a large database which requires longer time to transfer (e.g., days or a week) than the period of its invocation by cron (e.g., 1 day).
- Disk overflow by transfer is avoided.
- Even if *obiupdate* is suddenly killed during a database transfer, it is easy to resume the transfer. Actually, *obiupdate* checks incomplete databases (partly transferred databases) on the local disk, and attempts to utilize them as much as possible, instead of restarting the database transfer from the beginning.

- Completeness of transfer is checked by two files containing checksum and file-list in the database. Only the complete databases are deployed after the transfer (in OBIEnv, database deployment is done by reporting the database version in a node to the information server).
- Old and useless version of a database is automatically deleted. In case a node is dispatched some tasks, the deletion is delayed since the version is potentially used by the tasks. After the deletion, it is also reported to the information server.

3. PARALLEL AND PIPELINED TRANSFER BY OBIUPDATE

3.1. OBIEnv and Peer-to-Peer File Sharing Software

As mentioned in introduction, it is well known that a single-threaded file transfer between two computers is really inefficient since it can not fully utilize a network bandwidth. However, attempts to parallel transfer from primary site further escalate the flash crowd problem. This problem is well recognized in the field of Peer-to-Peer file sharing.[7,8] Aside from legal use of such programs, it is noticeable that some of them are successfully avoiding the problem by transferring parts of file(s) in parallel from different servers. For instance, eDonkey[9] and BitTorrent[10] accelerate the diffusion of a file by dividing it, transfer the divided parts in parallel, and a client which has a part begins to play a role of a server.

Framework of biological database transfer in OBIEnv shares some points with Peer-to-Peer file sharing. Firstly, a node plays both server and client roles in database transfer. Secondly, a new file is typically provided by only one server. For these reasons, in spite of the fact that OBIEnv has not been decentralized yet, we adopted the approach of parallel and pipelined transfer in the design and implementation of *obiupdate*. As to the automatic division of a single file, it is not essential in *obiupdate* since a typical biological database consists of 10 or more files.

3.2. Algorithm of Database Transfer by *Obiupdate*

The following steps show an overview of database transfer by *obiupdate*:

1. Based on DBLIST in .obienv.conf, it searches for the latest version of the databases to be updated, by querying to the information server.
2. If that exists, it starts to transfer the latest database. The fact that the node is now downloading the version of the database is reported to the information server. Using this information, other nodes which attempt to download the same version of the database can recognize the node as a candidate of servers.

3. Two files containing checksum and file-list in the database are first transferred by using rsync. In OBIEnv, databases are deployed and used at a directory /obienv/*DB name*/*DB version*, where the directory is opened to the download by rsync. Also, downloaded files are stored by using the same rule of directory path (the same databases with different versions never conflict). In this sense, a node can simultaneously play two roles (server and client) of database transfer.
4. Based on the file-list transferred in the previous step, it starts to transfer each files in the database. The fact that the node is now downloading a file from a server is reported to the information server, file by file. This information can be used to know the degree of parallel transfer as client and server in a node. Also, file-wise reporting is adopted for successful transfer of a file, which forms a list of downloaded files in a node. Based on this information, *obiupdate* can search a set of servers with a desired file. During the whole database transfer, a set of candidate server names is periodically updated by querying to the information server. If two or more servers are available for different files, *obiupdate* downloads them in parallel by using multiple execution of rsync (maximum number of parallel downloading is set to 4). On the other hand, if two or more servers are available for the same file, *obiupdate* chooses one of them by evaluating; the server's degree of parallel transfer; and the expected network distance between the node and the server (computed by NETPATH information).
5. After all the files have been transferred successfully, *obiupdate* reports the existence of the version of the database at the node.

3.3. Practical Questions

To answer the following practical questions, we conducted the database transfer experiments described in the next section.

- How the order of file transfer (e.g. dictionary order of file names) affects the performance?
- In cases where many nodes in a LAN (or under the same switch) are attempting to download the same version of a database, multiple transfers of the same file by such nodes from the server(s) located at remote site(s) should be suppressed since it is a waste of time and the broad bandwidth of LAN is not utilized effectively. In other words, if a node is downloading a file from a server at remote site, other nodes in the same LAN (or switch) should force themselves to delay the download of the same file until the node finish it and available as the fastest server to them. How it affects the performance?

- How the number and the granularity of files in a database affect the performance?
- How the WAN bandwidth of the primary server site affects the performance?

4. EXPERIMENTAL RESULTS

4.1. Databases

The databases used in the experiments are shown in Table 1. unigene has 50 entry files taken from various species (1.8MB - 499MB in size). Corresponding to each entry file, a FASTA format file and five BLAST format files with extensions nhr, nin, nsd, nsi, and nsq are also contained in unigene. On the contrary, nr-aa-5M, nr-aa-50M, and nr-aa-500M were generated by splitting the original FASTA format file taken from nr-aa database into relatively smaller files with the sizes 5MB, 50MB, and 500MB, respectively. BLAST index files with extensions phr, pin, pnd, pni, psd, psi, and psq are common to these three versions of nr-aa databases. Sizes of these index files vary from 80KB to 787MB.

Table 1. Statistics on the four databases used in the experiments.

ID	unigene	nr-aa-5M	nr-aa-50M	nr-aa-500M
database name	UniGene	nr-aa	nr-aa	nr-aa
version (YY:MM:DD)	2005:03:02	2005:03:25	2005:03:25	2005:03:25
# of entry files	50	0	0	0
# of BLAST index files	250	7	7	7
# of FASTA files	50	244	31	4
total # of files	350	251	38	11
max file size	499MB	787MB	787MB	787MB
average file size	9.7MB	11.9MB	78.7MB	272.1MB
total database size	3401MB	2993MB	2993MB	2993MB

4.2. Nodes

In the experiments, we used 68 nodes classified in three types (Table 2). Type-A nodes are scattered to 18 sites in Japan (see Figure 2). Type-B nodes have 1000Mbps network interfaces and connected to a 1000Mbps switch. Since the Tier 1 router shown in Figure 1 is also connected to the same switch, Type-B nodes can fully utilize the throughput of VPN network in OBIGrid. Type-C nodes are relatively slow computers with 100Mbps network interface and connected to a 100Mbps switch. In addition, GSC site is far from JAIST site where the Tier 1 router is located.

Table 2. Specification of the nodes used in the experiments.

type of nodes / spec etc.	Type-A	Type-B	Type-C
# of nodes	18	21	29
location of nodes	scattered	JAIST site	GSC site
# of CPUs per node	1	1 (physical) 2 (logical)	1
CPU type	Pentium 4	Pentium 4	Pentium III
CPU speed	2.4GHz	3.0GHz	1.4GHz
memory size	1GB / node	1GB / node	1GB / node

Table 3. Network speeds between jnis and other Type-A nodes.

direction of transfer / node (site) name	from jnis to the node (Mbps)	from the node to jnis (Mbps)
cirobi (CIR)	4.15	8.73
dnis (doshisha)	8.14	14.49
gnis2 (GSC)	2.68	6.52
hpq02 (hp)	4.20	0.70
ipabnis (IPAB)	4.96	10.35
knis (KYUSHU)	6.24	7.24
kona02 (titech-konalab)	7.84	14.55
ktgr0 (jaist-is)	53.37	52.40
mrin (mri)	5.85	15.30
obism220 (ism)	7.80	14.35
pluto (Qdai-okahonlab)	6.11	7.28
raicho (JST)	7.86	14.53
ryukyu02 (RYUKYU)	2.78	3.98
sand (mikilab-doshisha)	7.93	14.47
tok1 (tok)	6.48	7.50
tus01 (TUS)	7.06	10.91
valentine (MKI)	3.98	3.70

Table 3 shows typical network speeds measured by netperf between a Type-A node (jnis) located at JAIST site and other 17 Type-A nodes scattered in different sites. Since OBIGrid consists of VPN tunnels organized as star topology, a network speed in Table 3 represents the bandwidth of WAN at a site.

Fig. 3 is a screenshot of OBIEnv monitor. Current version of the monitor visualizes transfer status of each node in addition to task execution status.

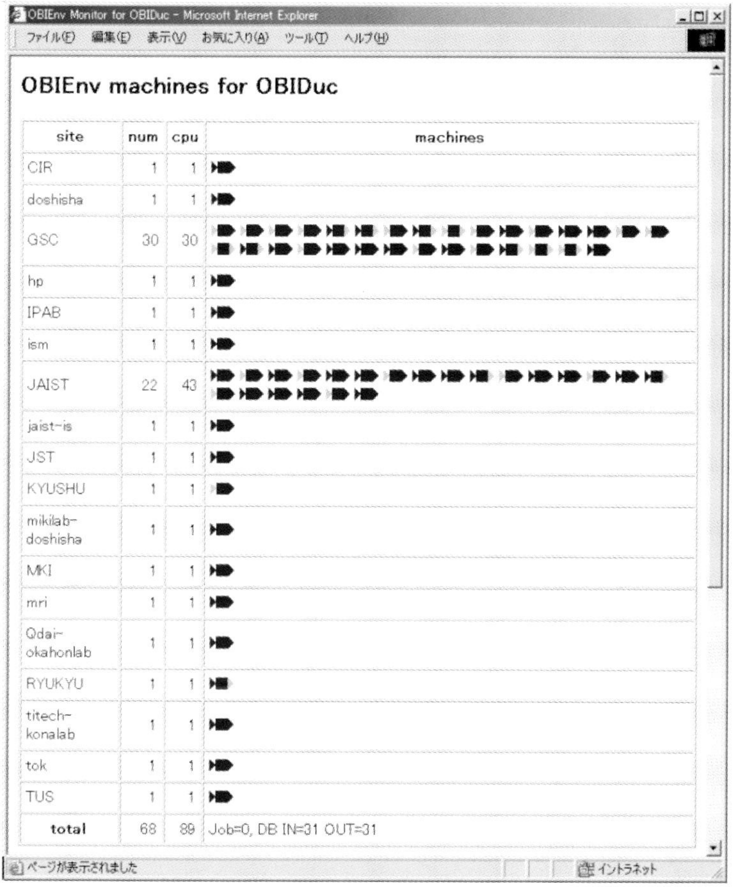

Figure 3. Screenshot of OBIEnv monitor. Each icon like a fish represents a node. A square indicates task execution status of a node, while two triangles adjacent to the square indicate transfer status (left for transfer-in, right for transfer-out). Light-gray triangles indicate active transfers.

4.3. Results

Using the databases and the nodes mentioned above, we conducted 9 experiments of database diffusion. The results of the experiments are illustrated in Figures 4-12. These figures show the percentage of transferred data size (i.e., 100% corresponds to the product of database size and the number of nodes (68)) and percentage of the number of nodes which completed to transfer whole database (i.e., 100% corresponds to 68 nodes).

In all of these experiments, a database was first deployed at a primary node. Then, other 67 nodes start to download the database by invoking *obiupdate* at the same time. Table 4 summarizes the differences among the experiments.

Table 4. Summary of the differences among the experiments.

Figure	DB	primary node	order of file transfer	suppress redundant transfers from the same LAN	finish time [HH:MM:SS]
Fig. 4	unigene	raicho	random	Yes	02:38:27
Fig. 5	unigene	raicho	dictionary order of file name	Yes	05:52:23
Fig. 6	unigene	raicho	ascending order of file size	Yes	06:39:30
Fig. 7	unigene	raicho	descending order of file size	Yes	07:55:50
Fig. 8	unigene	raicho	random	No	03:01:30
Fig. 9	unigene	ryukyu02	random	Yes	03:14:14
Fig. 10	nr-aa-500M	raicho	random	Yes	01:34:56
Fig. 11	nr-aa-50M	raicho	random	Yes	01:40:51
Fig. 12	nr-aa-5M	raicho	random	Yes	02:10:11

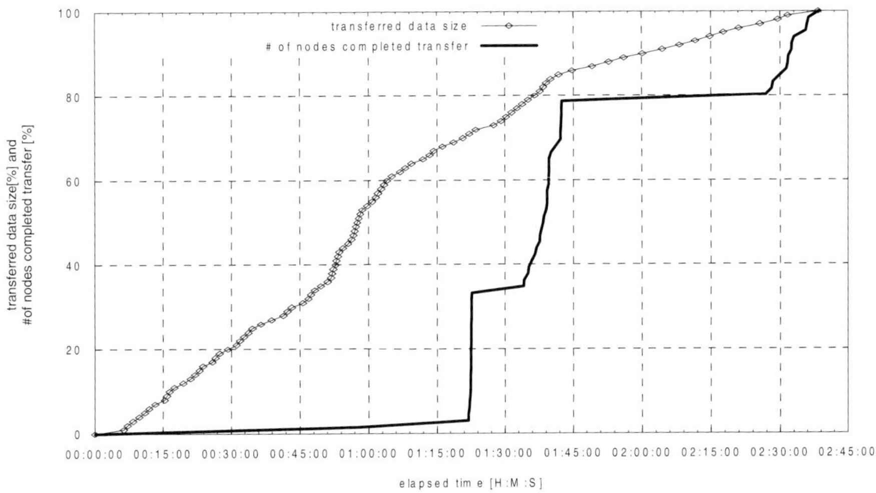

Figure 4. Graph of database diffusion. Database was unigene; primary node was raicho; random order transfer; redundant transfers from the same LAN were suppressed.

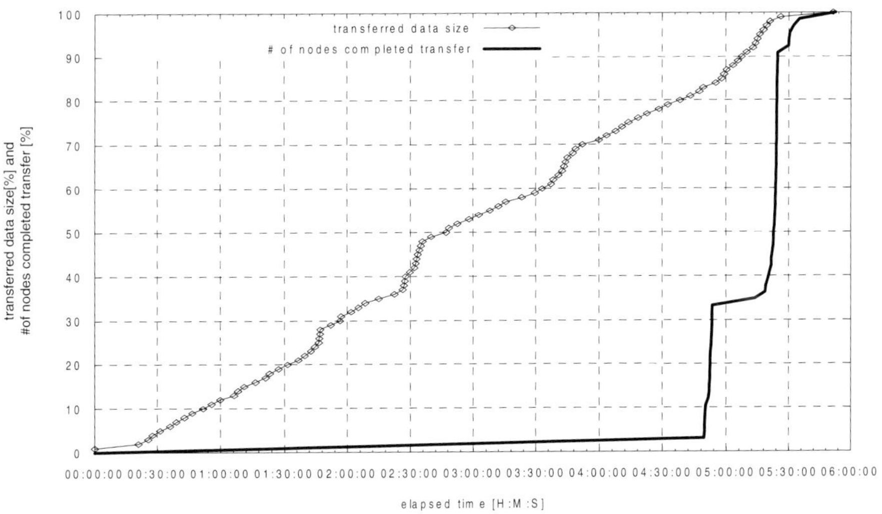

Figure 5. Graph of database diffusion. Database was unigene; primary node was raicho; transfer was done in dictionary order of file name; redundant transfers from the same LAN were suppressed.

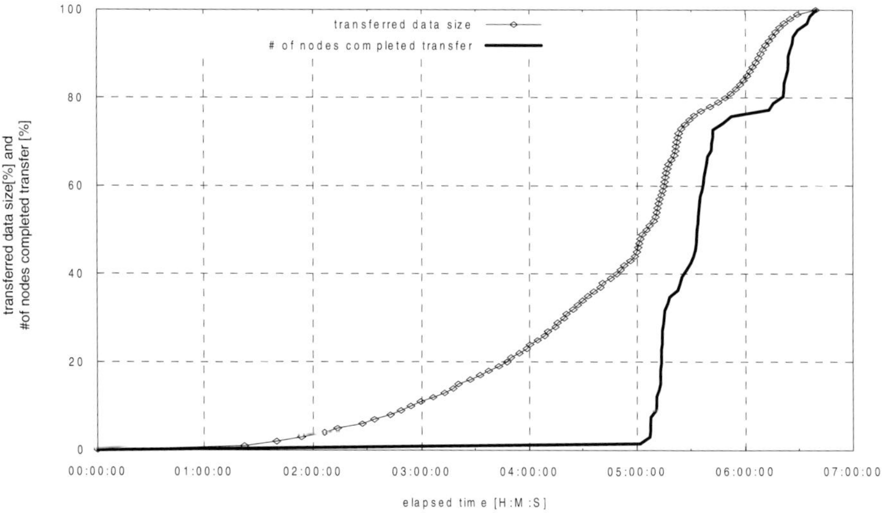

Figure 6. Graph of database diffusion. Database was unigene; primary node was raicho; transfer was done in ascending order of file size; redundant transfers from the same LAN were suppressed.

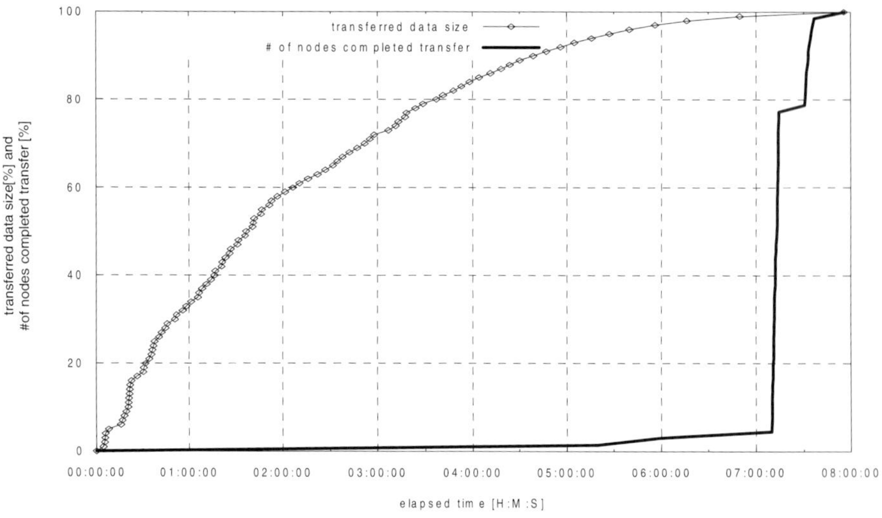

Figure 7. Graph of database diffusion. Database was unigene; primary node was raicho; transfer was done in descending order of file size; redundant transfers from the same LAN were suppressed.

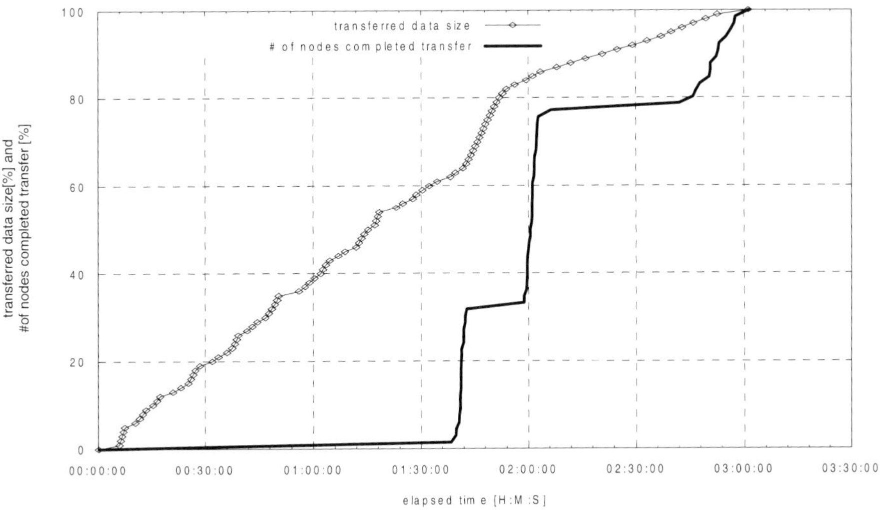

Figure 8. Graph of database diffusion. Database was unigene; primary node was raicho; random order transfer; redundant transfers from the same LAN were not suppressed.

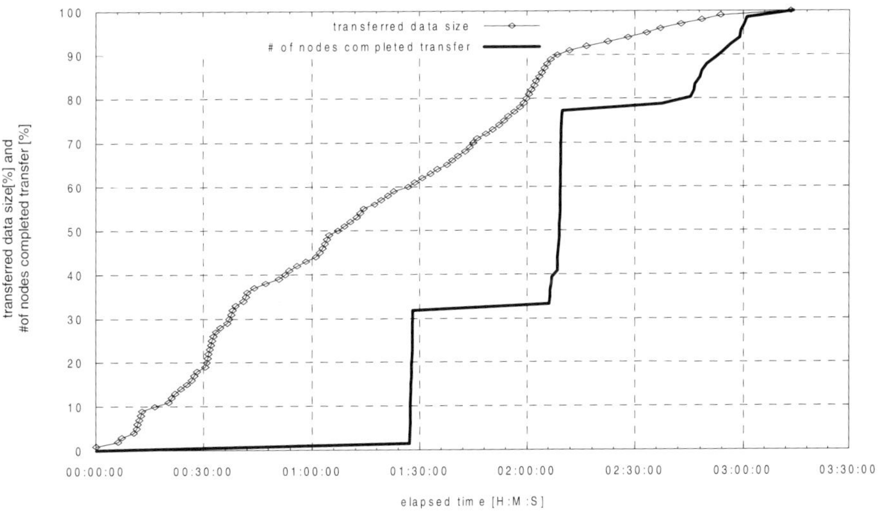

Figure 9. Graph of database diffusion. Database was unigene; primary node was ryukyu02; random order transfer; redundant transfers from the same LAN were suppressed.

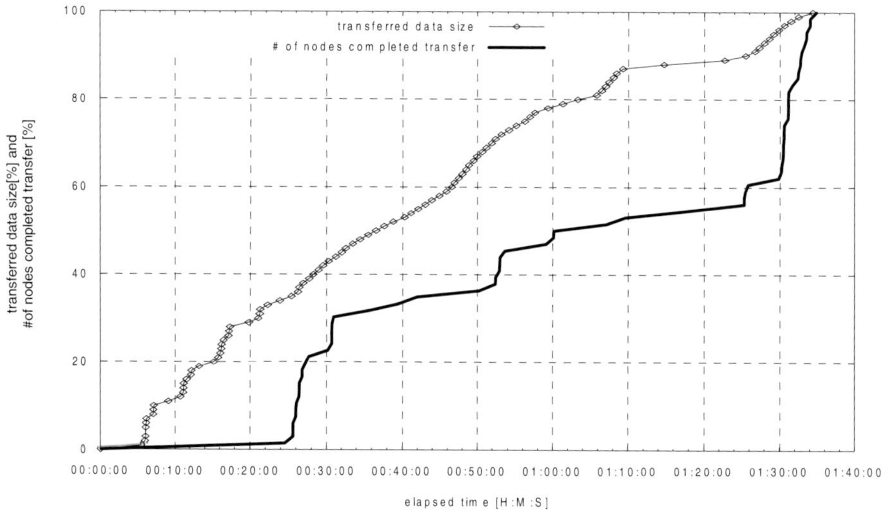

Figure 10. Graph of database diffusion. Database was nr-aa-500M; primary node was raicho; random order transfer; redundant transfers from the same LAN were suppressed.

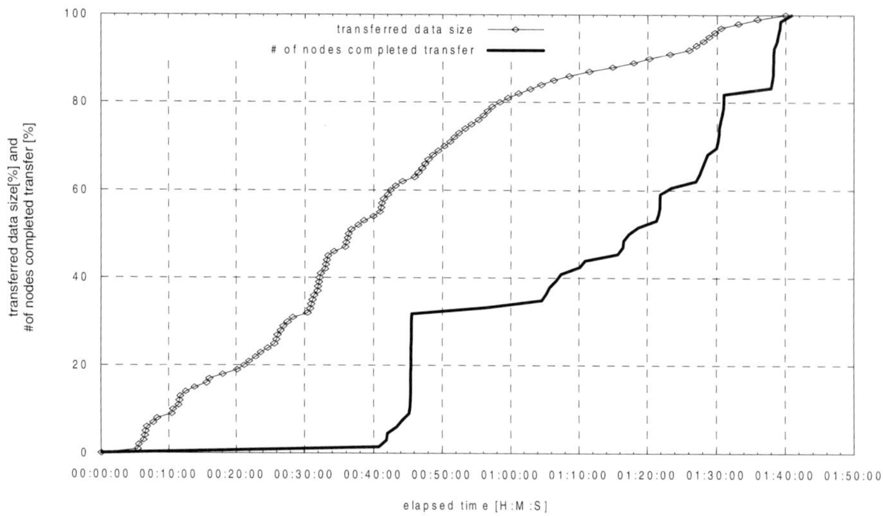

Figure 11. Graph of database diffusion. Database was nr-aa-50M; primary node was raicho; random order transfer; redundant transfers from the same LAN were suppressed.

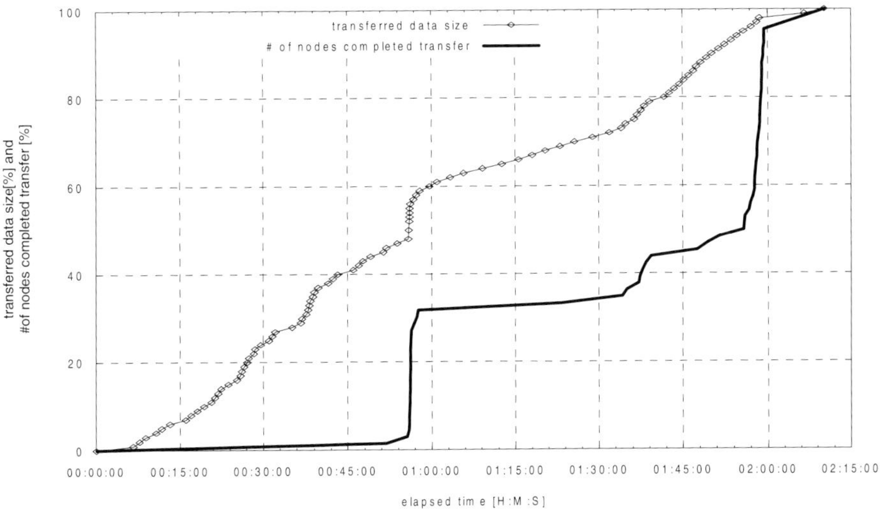

Figure 12. Graph of database diffusion. Database was nr-aa-5M; primary node was raicho; random order transfer; redundant transfers from the same LAN were suppressed.

4.4. Discussion

Current implementation of *obiupdate* was able to finish the transfer of unigene to 67 nodes within 02:38:27 (Fig. 4). Due to the existence of cluster nodes (Type-B and C), the number of nodes which completed the whole database transfer was doubled. Consequently, nearly 80% of nodes have completed the transfer at 01:42:00. On the contrary, in case we adopted a static order of file transfer, the diffusion of database was greatly slow (Figures 5–7). It is well deserved since a static order can worsen the performance of parallel and pipelined transfer of *obiupdate*. The increasing curves of transferred data size were different among Figures 5–7. However, it was common to them that completion of database transfer was delayed. In the worst case (Figure 7), only 8.95% of the nodes completed the transfer at 07:10:00. It means that most of nodes have kept having incomplete database for a long time.

In Figure 8, we can see the effect of suppressing redundant transfers from the same LAN. Comparing with Figure 4, finish time was delayed about 23 minutes. The tendency in the increase of the nodes which completed the transfer was very similar to the one in Figure 4, while the increasing curve of transferred database size was different. In Figure 4, from 00:51:00 to 01:05:00, we can see the first acceleration of the increase rate of transferred data size (the second is from 01:27:00 to

01:40:00). These two accelerations were caused by the existence of cluster nodes, however, they occurred at the same time or prior to the two jump-ups mentioned above. It means that in the case of Figure 4, cluster nodes effectively worked even if they have not completed the transfer. In Figure 8, the first acceleration occurred from 01:43:00 to 01:52:00 after the first jump-up at 01:40:00, and the second was not observed.

The only difference between the cases in Figures 4 and 9 is the choice of primary node. The outbound bandwidths of raicho and ryukyu02 were 14.53Mbps and 3.98Mbps, respectively (see Table 3). By choosing ryukyu02 as a primary node instead of raicho, the completion time was delayed about 36 minutes. Considering the fact that the single-threaded transfer of unigene from ryukyu02 to the center of OBIGrid (jnis) takes about 2 hours (3401MB * 8 / 3.98Mbps = 6836 sec. = 01:53:56) in addition to the fact that *obiupdate* uses at most 4 parallel rsync, it can be said that the 36-minute delay is reasonable. Furthermore, it is noticeable that in Figure 9, the first acceleration occurred from 00:30:00 to 00:42:00 prior to the first jump-up at 01:28:00.

Figures 10–12 illustrate the effect of the number and the granularity of files in a database. In these experiments, the completion time was delayed as many and smaller files are contained in a database. One of the possible reasons is the overload of the information server of OBIEnv. Since *obiupdate* adopts file-wise reporting to the information server, the total number of reporting is proportional to the number of files in a database. To solve the problem, we have to consider a decentralized approach of information server in OBIEnv in the future.

5. CONCLUSION

In this paper, we described the design and the implementation of *obiupdate*, which can efficiently spread a database to computers scattered at multiple sites by avoiding the problem so called flash crowds. Through the 9 experiments of database diffusion conducted on the real Internet environment, some performance features of *obiupdate* were shown; 1) random order transfer is two or three times faster than a transfer with static order; 2) *obiupdate* works well even if a primary node has relatively narrower outbound bandwidth; 3) massive files in a database makes the transfer slow, which might be caused by the overload of the centralized information server in OBIEnv. One possible way of improving the performance of *obiupdate* is to integrate sophisticated software for Peer-to-Peer file sharing into OBIEnv.

ACKNOWLEDGMENTS

This work was partially supported by a Grant-in-Aid for Scientific Priority Areas (C) "Genome Information Science" from the Ministry of Education, Culture, Sports, Science, and Technology of Japan.

REFERENCES

1. Satou S, Nakashima Y, Tsuji S, Defago X, Konagaya A. An Integrated System for Distributed Bioinformatics Environment on Grids. *Lecture Notes in Bioinformatics* 2005; **3370**; 8-19.
2. Konagaya A. OBIGrid: Towards a New Distributed Platform for Bioinformatics. *21st IEEE Symposium on Reliable Distributed Systems (SRDS'02)* 2002; 380-381.
3. Konagaya A, Konishi F, Hatakeyama M, Satou K. The Superstructure Towards Open Bioinformatics Grid. *Journal of New Generation Computing* 2004; **22**; **2**; 167-176.
4. OBIDuc. https://access.obigrid.org/jaist/obiduc/
5. Sugawara H. Gene Trek in Procaryote Space Powered by a GRID Environment. *Lecture Notes in Bioinformatics* 2005; **3370**; 1-7.
6. OBIEnv. https://access.obigrid.org/jaist/obienv/
7. Yang X, de Veciana G. Service Capacity of Peer to Peer Networks. *Proceedings of IEEE INFOCOM 2004* 2004; **4**; 2242-2252.
8. Stavrou A, Rubenstein D, Sahu S. A Lightweight, Robust P2P System to Handle Flash Crowds. *Proceedings of 10th IEEE International Conference on network Protocols (ICNP'02)* 2002; 226-235.
9. eDonkey. http://www.edonkey2000.com/
10. BitTorrent. http://www.bittorrent.com/

CONTROLLING THE CHAOS: DEVELOPING POST-GENOMIC GRID INFRASTRUCTURES

RICHARD SINNOTT AND MICHA BAYER[†]

National e-Science Centre || Bioinformatics Research Centre,
University of Glasgow, Glasgow G12 8QQ, Scotland
Email: ros@dcs.gla.ac.uk

"Why does Scotland have one of the highest rates of heart attacks in Europe? Are there genetic factors which contribute to this statistic"? The analysis and exploration of a broad array of life science data sets are needed to answer such questions. The Grid provides, at least conceptually, one way in which these kinds of data sets can be linked and analyzed. The life science domain places specific requirements on the Grid infrastructure needed to answer such questions. In this paper we describe these requirements and outline how they are being addressed in the DTI funded BRIDGES project.

1. INTRODUCTION

The life science community is experiencing a period of unprecedented change, challenge and opportunity. With the completion of the sequencing of the human genome (and ever increasing numbers of other genomes), the opportunities of *in-silico* scientific research offer a new horizon of possibilities: from rapid targeted drug discovery, identification of genetic factors to disease causes and epidemiological studies through to complete biological understanding of complete organisms and tailored genetic treatments supporting e-Health solutions. The possibilities abound! Fundamental to the realization of this vision is the infrastructure needed to use and understand the vast array of data sets associated with such research as depicted in Figure 1. These data sets are growing exponentially, have radically different characteristics, are often maintained by completely different groups and bodies, and importantly are perpetually evolving. In this context, the development of an infrastructure that allows to access, use, and

[†] Work supported by DTI grant to BRIDGES project.

analyze such changing and growing amounts of data is both technically challenging, offers huge benefits to the scientific community and is potentially extremely viable commercially.

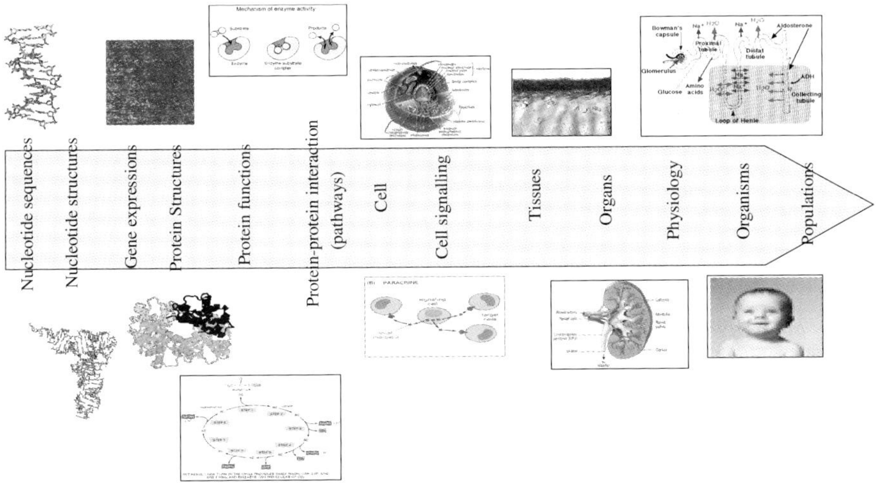

Figure 1. Spectrum of Life Science Data.

To support the vision of e-Health, it is clear that computational infrastructures must address (at least) the following requirements:

- Develop tools that allow simplified access to and usage of the potentially complex data structures that comprise life science data sets;
- Provide access to large scale computational resources needed to process and search the life science data sets, e.g. when comparing genomes;
- Ensure that appropriate security mechanisms are in place to deal with the data sets and infrastructure upon which they exist;
- Make this infrastructure easy to use and ideally targeted towards the needs of the specific scientific groups.

In the following sections we shall see how the BRIDGES project[1] is realizing these requirements through the development of a state of the art Grid infrastructure.

2. BIOMEDICAL RESEARCH INFORMATICS DELIVERED BY GRID ENABLED SERVICES

The Biomedical Research Informatics Delivered by Grid Enabled Services (BRIDGES) project[1] has been funded by the UK Department of Trade and Industry to develop a computational infrastructure to support the needs of the Wellcome Trust funded (£4.34M) Cardiovascular Functional Genomics (CFG) project[2]. The CFG consortium is investigating possible genetic causes of hypertension, one of the main causes of cardiovascular mortality. This consortium, which involves five UK and one Dutch site (depicted in Figure 2), is pursuing a strategy of combining studies on rodent models of disease (mouse and rat) contemporaneously with studies of patients and population DNA collections.

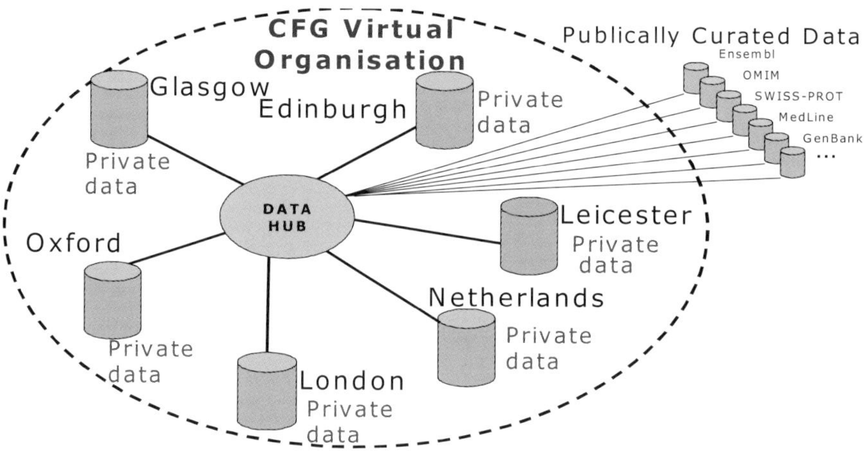

Figure 2. Data Distribution and Security of CFG Partners.

Currently many of the activities that the CFG scientists undertake in performing their research are done in a time consuming and largely non-automated manner. This is typified through "internet hopping" between numerous life science data sources. For example, a scientist might run a microarray experiment and identify a gene (or more likely set of genes) being differentially expressed. This gene is then used as the basis for querying a remote data source (e.g. MGI in Jackson[3]). Information retrieved from this query might include a link to another remote data source, e.g. on who has published a paper on this particular gene in MedLine[4] or PubMed[5]. Information from these repositories might include links to Ensembl[6] where further information on this gene, e.g., its start and end position in a given chromosome can be established. Such sequences of navigations typify the research undertaken by scientists.

2.1. Simplified Access To and Targeted Usage of Life Science Data Sets

A key component of the architecture in Figure 2 is the Data Hub. This represents both a local data repository, together with data made available via externally linked data sets. These data sets exist in different heterogeneous, remote locations with differing security requirements. Some data resources are held publicly (e.g. genome databases such as Ensembl[6], gene function databases such as OMIM[7] and relevant publications databases such as MedLine[4]); whilst others are for usage only by specific CFG project partners (e.g. microarray data sets[8] or quantitative trait loci (QTL) data sets[9]).

Currently the public data sets are accessible via two different technologies: IBM's Information Integrator (IBM-II) – formerly known as DiscoveryLink[10] (and soon to be known as Masala), and the Open Grid Service Architecture – Data Access and Integration (OGSA-DAI) technology[37]. IBM-II technology has been developed to meet the challenge of integrating and analyzing large quantities of diverse scientific data from a variety of life sciences domains offering single-query access to existing databases, applications and search engines. This is achieved through wrappers which use the data source's own client-server mechanism to interact with the sources in their native dialect. Through IBM-II access to a broad array of heterogeneous data sets can be achieved, e.g. in relational databases, XML databases, Excel spreadsheets, flat files etc.

In a similar vein, the OGSA-DAI technology provides a generic data access and integration mechanism overcoming issues related to the heterogeneity of technologies and data sets as well as the remoteness of data sets themselves. This technology is being continually refined and extended to be compliant with on-going Grid standardization efforts. At the time of writing, initial experiments are on-going in performing performance benchmarks against these two solutions for access to and usage of life science data.

To support the life science community, it is essential that applications are developed that allow them simplified access to life science data sets as well as to personalize their environments. The personalization might well include the data sources that are of most interest to the scientists and the information that they are most interested in from those data sources.

To support such personalization, the BRIDGES project has developed the application MagnaVista[1]. This application provides a completely configurable environment through which the scientists can navigate to and access a broad array of life science data sets of relevance to their research. The basic user interface to MagnaVista is depicted in Figure 3. Here the user can include the genes that they are most interested in (central pop up window). The lower left corner of Figure 3 lists the remote data sources that are accessible (SWISS-PROT[11], MGI[3], Ensembl[6] (rat, mouse, human DBs), RGD[12], OMIM[7]). The pop-up window to the right of Figure 3 shows the data sets that are returned to the user upon submission of the query.

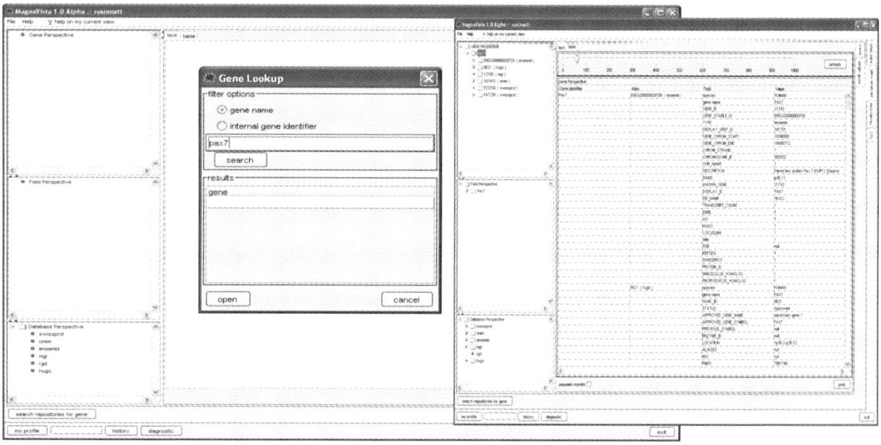

Figure 3. MagnaVista Basic Usage for Gene Query.

Thus rather than the user manually hopping to each of these remote resources, a single query is used to deliver collections of data associated with the genes of interest. To support the specific targeted data needs of the scientists, the MagnaVista application can be personalized in various ways. It currently supports user-based selection of specific (remote) databases that should be interrogated; user-based selection of the various data sets (fields) that should be returned from those databases; storage of specific genes of interest, as well as personalization of the look and feel of the application itself.

2.2. Tools Supporting Cognitive Aspects of Complex Data Sets

Life science data sets are notoriously complex, requiring a great deal of expertise to understand and utilise. Tools that facilitate cognitive understanding of these data sets, e.g., through visualisation are essential. One such relation that is especially important when dealing with translational studies between different genomes is *synteny*. Synteny is the condition of two or more genes being located on the same chromosome. Of particular importance is *conserved synteny* which may be defined as the condition where a syntenic group of genes from one species have orthologues and/or homologues in another species, i.e., similar sets of genes where the similarity itself can be ascertained through a combination of approaches such as protein sequence similarity, structure, and function.

The analysis of conserved synteny between the different organisms (e.g., mouse, rat and human), in combination with quantitative trait loci (QTL) data[9] and microarray experiments[8], is one of the main methods used by the CFG scientists in investigating hypertension. Their aim is to discover genes responsible for hypertension in rat or mouse organisms and translate these findings into knowledge about the mechanisms for hypertension in human. It should be noted that knowledge of syntenic relationships and of known QTLs between organisms provides supporting, but not necessarily guaranteed, evidence about the location and functional role of candidate genes causing hypertension between species. In displaying conserved synteny, two (or more!) chromosomes need to be shown simultaneously. SyntenyVista was developed for this purpose as shown in Figure 4.

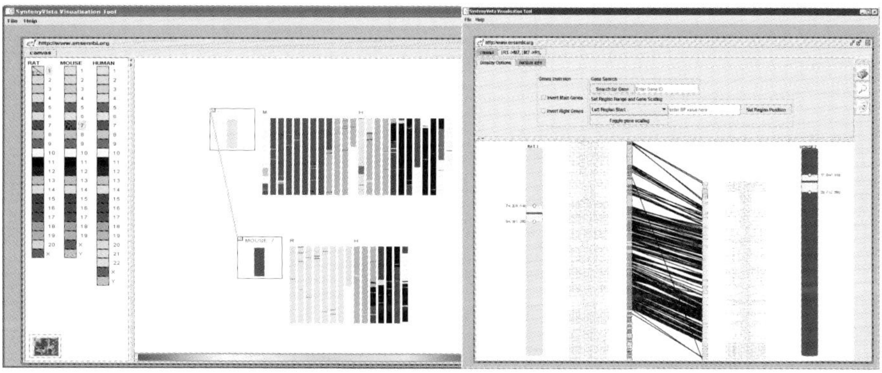

Figure 4. Grid Enabled Syntenic Visualization Tool.

The three columns of shaded boxes on the left of Figure 4 are the sets of chromosomes for the rat, mouse and human genomes. Users of SyntenyVista are able to drag these individual colour coded chromosomes onto the pallet where the QTL information is depicted. This represents the level of syntenic relations between rat, mouse and human chromosomes. In this example, the rat 1 and mouse 7 chromosomes are being visualized since they share a high degree of synteny. Users are then able to show detailed information on the relationships between the many thousands of genes in these two chromosomes as depicted on the right of Figure 4. Further information is also available to the users when scrolling through these chromosomes. For example, users are able to pan out to visualize the complete chromosomes, pan in to see individual relationships between specific genes, find where specific genes start and end on the chromosome, and gain access to detailed information associated with these genes.

2.3. Services for Simplified Usage of Large Scale Computational Resources

In their pursuit of novel genes and understanding their associated function life scientists often require access to large scale compute facilities to analyse their data sets, e.g., in performing large scale sequence comparisons or cross-correlations between large biological data sources. The Basic Local Alignment Search Tool (BLAST) [13] has been developed to perform this function. Numerous versions of BLAST currently exist which are targeted towards different sequence data sets and offer various levels of performance and accuracy metrics. BLAST involves sequence similarity searches, often on a very large scale, with query sequences being compared to several million target sequences to compute alignments of nucleic acid or protein sequences with the goal of finding the *n* closest matches in a target data set. BLAST takes a heuristic (rule-of-thumb) approach to a computationally highly intensive problem and is one of the fastest sequence comparison algorithms available.

There are a number of public sites[14,-15] that provide users with web-based access to BLAST, and these generally use dedicated clusters to provide the service. However, the growth of sequence data over the last decade has been exponential[16], and consequently searches take increasingly longer. Given that most biologists use these public sites rather than running the computation on their own machines (BLAST is freely available as a command line tool), the load on the purpose built clusters has increased dramatically and now significant queuing times are becoming increasingly common. A typical use of BLAST will usually involve searching against the *nt* database from NCBI[17] - a data set that contains several of the main sequence repositories and is currently ~3 GB in size (with over 2,2M target sequences).

Usage of BLAST on large scale HPC resources is often non-trivial to achieve and typically requires knowledge of scripting languages (for decomposing the input data sets and recomposing/merging the results data) and local job schedulers. Users should not have to learn the often complex options associated with job submission to job schedulers such as Condor[18] or OpenPBS[19]. In addition, one of the primary benefits of Grid technology is the ability to dynamically select and use a variety of heterogeneous resources is essential. This in turn requires that meta-schedulers are available that can dynamically schedule jobs across a variety of heterogeneous resources utilising a variety of local job schedulers. The BRIDGES Grid BLAST service which provides such a simplified BLAST based job submission system, enabling access to and usage a collection of HPC facilities is shown on the left of Figure 5.

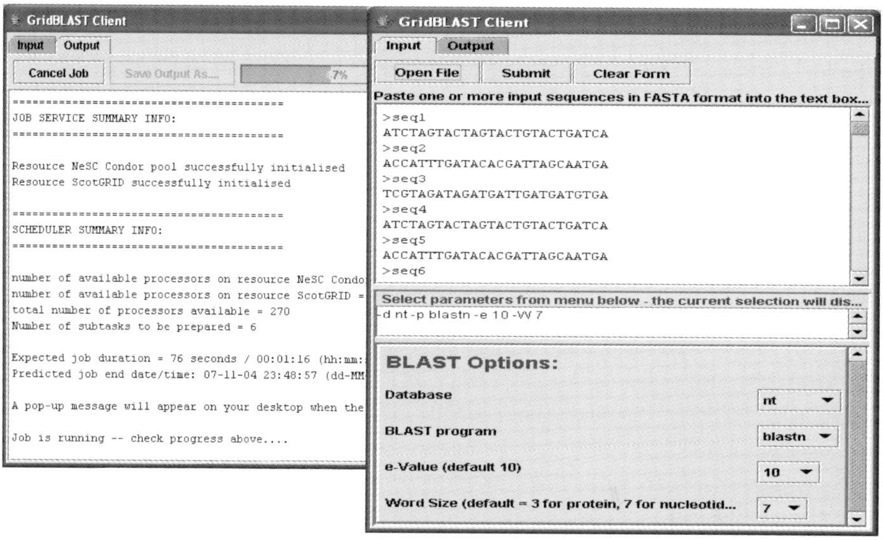

Figure 5. Grid enabled BLAST service and Monitoring its Usage.

An explicit requirement in designing this service was that it should be extensible. This has been achieved through XML based resource configuration files which easily allow new sets of resources to be added and subsequently used. Currently this Globus toolkit version 3[20] service provides access to the ScotGrid computational resource[21] and a Condor pool at the University of Glasgow. The ScotGrid resource is *the* e-Science resource at the University of Glasgow and represents a consolidation of resources across a variety of research groups and departments. It consists of the equivalent of 255 1GHz processors (using hyper-threading) and 15TB disk space comprised of IBM xSeries, Blade server, FAStT500 and Dell and Cisco technologies. It uses the Maui scheduling software[22] which implements the scheduling policies for the OpenPBS[19] batch submission system.

This service provides intelligent default settings for a variety of BLAST services. When used, the service checks what resources are available, where the jobs are best run and subsequently provides a prediction of how long the complete BLAST job will take to complete. In addition, monitoring of the status of the various sub-jobs is undertaken and staging of the various input and output files onto the compute resources is provided. This is indicated on the right of Figure 5. At the time of writing, the BLAST service is being extended to make use of the UK e-Science National Grid Service[41].

2.4. Supporting a Scalable and Fine Grained Security Infrastructure

The widespread acceptance and uptake of Grid technology can only be achieved if it can be ensured that the security mechanisms needed to support Grid based collaborations are at least as strong as local security mechanisms. The predominant way in which security is currently addressed in the Grid community is through Public Key Infrastructures (PKI) [23] to support *authentication*. Whilst PKIs address user identity issues, authentication does not provide fine grained control over what users are allowed to do on remote resources (*authorization*). The Grid community has put forward numerous software proposals for authorization infrastructures [24-26]. It is clear that for the foreseeable future a collection of solutions will be the norm.

Key concepts associated with authorization are a Policy Enforcement Point (PEP) and a Policy Decision Point (PDP). The PEP ensures that all requests to access the target are authorized through checking with the PDP. The PDP's authorization decision policy is often represented through collections of rules (policies). The different authorization infrastructures associated with Grid technology have put forward their own mechanisms for realizing PEPs and PDPs. Recently however the Grid community has put forward a generic API – the SAML AuthZ API[27]. This is an enhanced profile of the OASIS Security Assertion Markup Language v1.1[28].

The OASIS SAML AuthZ specification defines a message exchange between a policy enforcement point (PEP) and a policy decision point (PDP) consisting of an *AuthorizationDecisionQuery* flowing from the PEP to the PDP, with an assertion returned containing some number of *AuthorizationDecisionStatements*. The *AuthorizationDecisionQuery* itself consists of: a *Subject* element containing a *NameIdentifier* specifying the initiator identity; a *Resource* element specifying the resource to which the request to be authorized is being made, and one or more *Action* elements specifying the actions being requested on the resources.

The GGF SAML profile specifies a *SimpleAuthorizationDecisionStatement* (essentially a granted/denied Boolean) and an *ExtendedAuthorizationDecisionQuery* that allows the PEP to specify whether the simple or full authorization decision is to be returned. In addition the GGF query supports both the pull and push modes of operation for the PDP to obtain attribute certificates, and has added a *SubjectAttributeReferenceAdvice* element to allow the PEP to inform the PDP where it may obtain the subject's attribute certificates from.

Through this SAML AuthZ API, a generic PEP can be achieved which can be associated with arbitrary (GT3.3+) Grid services*. Thus rather than developers having to explicitly engineer a PEP on a per application basis, the information contained within the deployment descriptor file (.wsdd) when the service is deployed within the container, is used. Authorization checks on users attempting to invoke "***methods***" associated with this service are then made using the information in the .wsdd file and the contents of the LDAP repository (PDP) together with the DN of the user themselves. Note that this "method" authorization basis extends current security mechanisms such as GSI which work on a per service/container basis. This generic solution can be applied to numerous infrastructures used to realize PDPs such as PERMIS.

PERMIS provides a Role Based Access Control (RBAC) infrastructure. RBAC models have been designed to make access control manageable and scalable[29]. PERMIS provides a standards-based Java API that allows developers of resource gateways to enquire if a particular access to the resource should be allowed. PERMIS provides tools that allow creation of XML based policies defining rules, specifying which access control decisions are to be made for given resources, e.g. definitions of subjects that can be assigned roles; definitions of Source of Authority (SOA) - trusted to assign roles to subjects; definitions of roles and their hierarchical relationships; definitions of what roles can be assigned to which subjects; definitions of targets that are governed by the policy, and the conditions under which a subject can be granted access.

Both PERMIS and the Globus toolkit version 3.3 (GT3.3) have been extended to support the SAML AuthZ API. PERMIS tools such as the Policy Editor and Privilege Allocator have been applied to create XML based policies which allow restricted access to the compute and data resources to the various CFG scientist roles.

The work on BRIDGES has applied this authorization infrastructure to restrict (authorize) access to specific data sets within the federated repository and to the

---

* It has been stated by the Globus team (V. Welch, J. Schopf) that this API will be supported in future versions of the Globus toolkit.

specific compute resources that are accessible via the BLAST service. The current resource specific policy that is supported is based upon three key roles depending upon whether the user has a valid UK e-Science certificate; and/or whether the user has a local account on the HPC facility (ScotGrid) at Glasgow. If neither of these conditions is true, then the user may only perform a BLAST job on the freely available Condor pool at Glasgow University. This has been demonstrated to work, however the scalability of such low level policies is an issue that must be resolved.

2.5. Portal Technologies and Simplified Delivery Mechanisms

There are various possibilities available for hosting the services to be made available to the CFG scientists. Given that user friendliness is a key aspect, development of a project portal was made. This portal provides a personalizable environment that the scientist is offered to explore all of the (Grid related) software, data resources and general information associated with the BRIDGES, and hence the CFG projects.

Arguably the most mature portal technology on the market and the market leader is IBM WebSphere Portal Server[30], which has been used to develop the BRIDGES portal, although we note that several other solutions were also investigated, including GridSphere[31] and the Commodity Grid toolkit[32]. WebSphere Portal Server runs as another layer on top of the highly developed WebSphere Application Server. Since this provides a fully functional enterprise Java hosting environment it is possible to deploy a Java based Grid service instance within the same virtual machine container.

The BRIDGES portal itself provides an integrated and personalizable environment through which the scientists have access to the various Grid services that they need. This includes the MagnaVista service, the SyntenyVista service, the Grid BLAST service and other services allowing the scientists to store and share a variety of bioinformatics data sets, including Quantitive Trait Loci (QTL) and microarray data sets (based upon the MIAME compliant services). Depending upon their role within the project, the personalization of the portal to the scientist is based upon secured (authorized) profiles accessible via the GGF SAML AuthZ API.

Simplified delivery mechanisms are crucial to ensure the success of Grid based technologies to the wider community. It is infeasible to expect non-computer scientists to have to deal with software deployment aspects related to the set up and configuration of the complex infrastructures associated with Grid technology. To address this issue, one mechanism that has been successfully explored within the BRIDGES project is Sun's WebStart technology[33]. Through this technology, users require only a browser to gain access to the various services. The portal provides *launch* buttons which when selected by the end user, check whether WebStart technology exist on the remote (end user) system. If this is not the case the user is prompted if they want to install this and if so, it is automatically installed along with the application itself. WebStart also allows easy handling of changes and updates to the Grid services available from the portal itself through providing checks on the latest version of the applications available on the end user systems.

3. CONCLUSIONS

The BRIDGES project began in October 2003 and investigated a wide variety of Grid technologies applicable to the life science domain. The current implementation status has provided a proof of concept prototype. Grid technologies *do* allow for simplified access to and usage of a broad set of post-genomic data sets – bringing the data to the scientist! Services to support the analysis and visualization of these large life science data sets, efficiently utilizing HPC facilities have also been realized, taking into consideration appropriate security mechanisms deemed applicable. In short, it works! The work has not been without issues. The stability of the Grid middleware such as the Globus toolkit and the associated documentation remains below an appropriate level to easily produce Grid based systems. Compromises had to be made between the architecture and design, and the final systems that have been implemented due to for example, operating system dependencies of the middleware.

The work is evolving based upon feedback from usage of the infrastructure by the CFG scientists. Close liaison with the scientific community is essential to ensure that we are developing the "right software" and accessing the right data sets. Given that the CFG project are primarily interested in functional genomic based data resources, e.g., in supporting their microarray experiments, a bioinformatics workbench that allows the scientists to take up/down regulated gene names from microarray experiments and garner further information are of special interest. We note that the data sets accessible via the Data Hub are not a fixed set. Other resources can easily be added. However one of the challenges in this area is the issue in gaining access to these remote resources. For example, few of these resources provide the necessary programmatic access needed, i.e., to query their database directly. Instead, they often only offer compressed files that can be downloaded. As a result, the Data Hub includes a local warehouse for downloaded data sets. Currently federated access to Ensembl (rat, mouse, human) and MGI is supported, with the other data sets having to be locally warehoused. This then requires that local (downloaded) data sets are kept up to date with the remote data sets.

In our experience it is often non-trivial to establish a local repository that uses theses data sets, even with the local downloading of the data. Thus for example, the data providers often do not provide schemas or data models for the data themselves. Rather they simply provide large compressed text files that can be FTP-ed. It requires a significant amount of effort to understand how this data can be input into a database, i.e., in working out what associated the data model/schema is. To address this, numerous funding councils in the UK (MRC, BBSRC, NERC, JISC, DTI, Wellcome Trust) have funded the Joint Data Standards Survey project [34]. This is investigating the technical, social, political and often ethical issues that are currently prohibiting the effective sharing of life science data sets, with the idea being that policies can be formulated to facilitate life science data sharing.

Underpinning data sharing is of course agreeing upon (standardizing) data models and technologies used to access these resources. The Grid community[35] is currently defining standards[36] for access to data on the Grid. The OGSA-DAI project[37] in particular has helped to shape this work and has produced Grid based implementations. A report comparing OGSA-DAI and IBM Information Integrator is currently under production.

Access to a broad range of life science data sets is essential if the vision of a future e-Health infrastructure is to be achieved. As described, the BRIDGES project has focused on the left hand side of Figure 1: functional genomics. Extending this to incorporate a wider variety of data resources and incorporating further life science applications is an on-going process. Two new projects at Glasgow in particular will allow for this process to be explored. The 4-year Scottish Bioinformatics Research Network (SBRN) project[38] will allow to both expand the number of applications available within BRIDGES, and to bring the existing prototypes to a more robust, production quality level. The 3-year Virtual Organisations for Trials and Epidemiological Studies (VOTES) project[39] is exploring how Grid based technologies can be applied in the clinical trials domain. Clinical trials require up to date information on patterns of disease/frequency of clinical procedures. VOTES will explore three aspects of clinical trials: recruitment of patients; data collection on those patients; and tools that facilitate the management of a clinical trial. Since the information that underpins clinical trials is located in a range of highly secure sites such as GP/doctor databases, hospital databases, clinical registries, death registries etc., exploring the applicability of the Grid in this domain will require extremely rigorous security and ethical considerations to be visibly supported. Linkage of such data sets with other genomics related data sets *a la* BRIDGES is necessary if e-Health is to be supported however, and to answer the opening question in the abstract!

ACKNOWLEDGEMENTS

This work was supported by a grant from the Department of Trade and Industry. The authors would also like to thank members of the BRIDGES and CFG team including Prof. David Gilbert, Prof. Malcolm Atkinson, Dr. Dave Berry, Dr. Ela Hunt and Dr. Neil Hanlon. Dr. Hanlon and Dr. Hunt are also acknowledged for their contribution to the original SyntenyVista software. Magnus Ferrier is acknowledged for his contribution to the MagnaVista software and Derek Houghton for his work in developing the data repository. Acknowledgements are also given to the IBM collaborators on BRIDGES including Dr. Andy Knox, Dr. Colin Henderson and Dr. David White. The CFG project is supported by a grant from the Wellcome Trust foundation.

REFERENCES

1. BioMedical Research Informatics Delivered by Grid Enabled Services (BRIDGES), www.brc.dcs.gla.ac.uk/projects/bridges
2. Cardiovascular Functional Genomics project, http://www.brc.dcs.gla.ac.uk/projects/cfg/
3. Mouse Genome Informatics (MGI), http://www.informatics.jax.org/
4. US National Library of Medicine, http://www.nlm.nih.gov/
5. PubMed Central Home, http://www.pubmedcentral.nih.gov/
6. EMBL-EBI European Bioinformatics Institute, http://www.ebi.ac.uk/
7. NCBI Online Mendelian Inheritance in Man, http://www.ncbi.nlm.nih.gov/OMIM/
8. Minimal Information About a Microarray Experiment (MIAME) http://www.mged.org/Workgroups/MIAME/miame.html
9. An Overview of Methods for Quantitative Trait Loci (QTL) Mapping, Lab of Statistical Genetics, Hallym University http://bric.postech.ac.kr/webzine/content/review/indivi/2002/Aug/1_08_index.html
10. IBM Information Integrator, http://www3.ibm.com/solutions/lifesciences/solutions/Information Integrator.html
11. SWISS-PROT, http://us.expasy.org/sprot/
12. Rat Genome Database, http://rgd.mcw.edu/
13. Basic Local Alignment Search Tool (BLAST), http://www.ncbi.nlm.nih.gov/Tools/
14. EBI BLAST website, http://www.ebi.ac.uk/blastall/index.html
15. NCBI BLAST website, http://www.ncbi.nlm.nih.gov/BLAST/
16. GenBank statistics web page, http://www.ncbi.nih.gov/Genbank/genbankstats.html
17. NCBI Nucleotide database, http://www.ncbi.nlm.nih.gov/entrez/query.fcgi?db=Nucleotide
18. Condor website, http://www.cs.wisc.edu/condor/

19. Open Portable Batch System, www.openpbs.org
20. Globus toolkit, www.globus.org/toolkit
21. ScotGrid, www.scotgrid.ac.uk
22. Maui Job Scheduler, http://www.supercluster.org/maui/
23. C Adams and S Lloyd, Understanding Public-Key Infrastructure: Concepts, Standards, and Deployment Considerations, Macmillan Technical Publishing, 1999.
24. Privilege and Role Management Infrastructure Standards Validation project www.permis.org
25. L Pearlman, et al., A Community Authorisation Service for Group Collaboration, in Proceedings of the IEEE 3rd International Workshop on Policies for Distributed Systems and Networks, 2002.
26. M Thompson, et al., Certificate-Based Access Control for Widely Distributed Resources, in Proceedings of the 8th Usenix Security Symposium, Washington, D.C., 1999.
27. V. Welch, F. Siebenlist, D. Chadwick, S. Meder, L. Pearlman, Use of SAML for OGSA Authorization, June 2004, https://forge.gridforum.org/projects/ogsa-authz
28. OASIS. Assertions and Protocol for the OASIS Security Assertion Markup Language (SAML) v1.1, 2 September 2003, http://www.oasis-open.org/committees/security/
29. D. Chadwick and A. Otenko. The PERMIS X.509 role based privilege management infrastructure, in Proceedings of the Seventh ACM Symposium on Access Control Models and Technologies, Monterey, California, USA, 2002.
30. WebSphere Portal software, http://www-306.ibm.com/software/genservers/portal/
31. GridSphere Portal, www.gridsphere.org
32. Commodity Grid toolkit, www-unix.globus.org/cog
33. Sun WebStart Technology, http://java.sun.com/products/javawebstart/
34. Joint Data Standards Survey Project, www.nesc.ac.uk/hub/projects/jdss
35. Data Access and Integration Services working group, https://forge.gridforum.org/projects/dais-wg
36. Grid Data Service Specification, http://forge.gridforum.org/docman2/ViewCategory.php?group_id=49&category_id=517
37. OGSA-DAI project, www.ogsadai.org.uk
38. Scottish Bioinformatics Research Network (SBRN), www.nesc.ac.uk/hub/projects/sbrn
39. Virtual Organizations for Trials and Epidemiological Studies (VOTES) project, www.nesc.ac.uk/hub/projects/votes
40. R. Sinnott, Grid Based Clinical Trials Scenarios, presentation given at Global Grid Forum Life Science Research Group, Brussels, September 2004, http://www.nesc.ac.uk/talks/ros/ClinicalTrialsOutlineScenariosv2.pdf
41. UK e-Science National Grid Services, www.ngs.ac.uk

DO GRID TECHNOLOGIES HELP LIFE SCIENCES?
LESSONS LEARNT FROM BIOGRID PROJECT IN JAPAN*

SUSUMU DATE†

Graduate School of Information Science & Technology, Osaka University, Japan

KAZUTOSHI FUJIKAWA

Information Technology Center, Nara Institute of Science & Technology, Japan

HIDEO MATSUDA

Graduate School of Information Science & Technology, Osaka University, Japan

HARUKI NAKAMURA

The Institute for Protein Research, Osaka University, Japan

SHINJI SHIMOJO

The Cybermedia Center, Osaka University, Japan

Biogrid project is one of national R&D projects on grid in IT-Program granted by Ministry of Education, Culture, Sports, Science and Technology since 2002. In this project, we consider a grid technology as a glue or a middleware to integrate observation devices, databases and computational resources for advanced life science. We have built four typical applications in life science as exemplars on grid to show the power and effectiveness of the grid middleware, i.e., protein prediction, multi-scale bio-simulation, and database federation. In addition to the current grid middleware such as Globus, the secure Grid file system and IPv6 based grid middleware form a foundation of these advanced applications. From 2005, we will tightly collaborate with NAREGI project, which is a national R&D grid middleware project stared in 2003 in terms of the development of simulation platform and data grid middleware.

Keywords: Grid, life scientific database federation, BioPfuga, Multi-scale simulation, OGSA

---

* Correspondence to: Susumu Date, PhD, s.date@ucsd.edu, The Center for Research in Biological Systems, University of California San Diego, 9500 Gilman Drive, La Jolla, CA 92093-0043

† Also The Center for Research in Biological Systems, University of California.

1. INTRODUCTION

A recent paradigm shift brought about by genome sequence analyses of many different species is about to dramatically change not only biological research but also everything involved in life science. Examples of such changes are partly characterized by the terms 'tailor-made medicine' and 'genome drug discovery'. Advanced life science requires a diversity of bio-related data to be effectively and efficiently treated.

Seamless access to heterogeneous and geographically distributed databases will take on a role of importance for life science research. At the same time, as the mechanism of biological organism reveals in cell or molecule level, the computational requirements for simulating biological phenomena increases. Also, as the observation devices in life sciences such as microscope, CT and fMRI, make progress in its preciseness and complexity, the requirements for data storage and network bandwidth for exchanging data grow more demanding. There needs some secure mechanism to share patient related data or drug candidate.

For the advanced life science, various elements of information communication technology are required such as database, security, computational science and communication. Grid seems to provide good solutions to the advanced life science in the integrated manner. Computing grid provides a way to integrate distributed computers and makes a large scale biological simulation possible through a huge computational power. Data grid provides a way to federate distributed heterogeneous biological databases and enables a single query to handle complicated biological relation. Workflow tools make it possible to perform trial-and-error procedure for predicting a complicated protein structure. However, we need collaboration between ICT researchers and life science researchers to clarify whether current grid technology will provide enough technology for the advanced life science and what the gaps of the current grid technology are.

Biogrid project was started in 2002 with this mission. By the collaboration of very generous and open-minded ICT and life science researchers we tried to solve biological problem by using grid technology. Through the careful dialogue among researchers, we focused on targeted applications in life science and 'gridified', found appropriate grid technology and applied them. These four applications are HTC (High Throughput Computing) for Ab initio protein structure prediction, BioPfuga, a Grid Platform for Bio-simulation, federation of Bio-databases, tele-science for Analysis workbench for remote observations and secure data sharing for life science. In this paper, we show three examples.

2. AB INISIO PROTEIN STRUCTURE PREDICTION ON GRID

In this section, we describe how high throughput computing technologies supports the protein structure prediction problem.

2.1. Protein Structure Prediction Server ``ROKKY"

HTC group uses a protein structure prediction system called "ROKKY" which is originally designed at Kobe University. ROKKY is a web-based fully-automated server that can predict a protein structure given an amino acid sequence. Performance of ROKKY was benchmarked in recent world-wide blind test at the CASP6 [1] showed that ROKKY is the second best prediction server among over 50 servers for targets without template protein. The distinctive feature of ROKKY is that the system integrates the standard bioinformatics tools and a fragment assembly simulator called "SimFold" [2, 3].

For all targets, ROKKY first performs sequence analysis with "PSI-BLAST" [4] using a non-redundant sequence database and the PDB. If an appropriate structural template is not found, ROKKY uses "3D-Jury" [5] for seeking available structural templates.

If any suitable template is found with both tools, ROKKY performs Fragment Assembly Simulated Annealing (FASA) with SimFold [6]. Fully-automated predictions, no matter how good they are in an assessment ranking, often fail in prediction for many unobvious reasons. Sometimes a visual inspection of early stage results by experts could reveal the reason of failure and the way for overcoming it. For some targets, experts know that FASA can predict more appropriate structure than a template-based modeling does. Thus, in some case, a human needs to intervene in the job execution of ROKKY. As a result, more informative results of ROKKY help future structure predictions, and these results become a good training set to learn the knowledge of experts.

2.2. Requirements for Protein Structure Prediction

Since a complete protein structure prediction method has not been established, bioinformaticians might use some prediction methods in combination and predict the structure through trial-and-error. They frequently need to change the input parameters of a protein structure prediction method or even protein structure prediction method itself.

However, they still define the execution order of methods and input parameters in the form of a batch script by using some script language such as Perl or shell script. Moreover, they cannot easily change the execution order of methods, because the description of a batch script is likely to be complicated. The execution of methods by a batch script is unsuitable for protein structure prediction, because each job of prediction methods requires enormous time to acquire the complete result. To solve such problem, an interactive application interface is required on which users can inspect arbitrary stage results and dynamically modify the input parameters or even the prediction methods.

2.3. Workflow Design and Control Tool

HTC group has been implementing a *workflow design and control tool* for users to easily define the execution order of several methods and input parameters of each method. From now on, we call a protein structure prediction method and the execution order of methods, "*work*" and "*workflow*," respectively.

Our workflow design and control tool provides a differential execution mechanism of a workflow. On this mechanism, users can freely modify a workflow and re-execute the workflow at an arbitrary point. We consider that the differential execution of the workflow is required in the following cases:

- Users find that the definition of a *workflow*, which is in execution is wrong and want to change it dynamically.
- Users want to change the *work*.
- Users want to change the input parameters of each *work*.

ROKKY uses "PSI-BLAST," "3D-Jury" and "FASA with SimFold." Each job executed in these three methods is treated as "work" in our tool.

Since the characteristic of each *work* varies, our tool assigns computational nodes for each *work* as follows: For both PSI-BLAST and 3D-Jury, a single node will be dispatched. For FASA with SimFold, available nodes will be dispatched, since SimFold executes a lot of tasks, each of which has different input parameters. The number of required nodes can be specified in a workflow. If the number of available nodes is less than the required one, our tool assigns nodes according to the strategy specified in the workflow.

Our workflow design and control tool provides a GUI (graphical user interface) for users to easily define a workflow (see Figure 1). On this GUI, users can perform the following operations: defining/modifying of a *work*, defining/modifying of a *workflow*, specifying/modifying of input parameters of a *work*, submitting of a *workflow*, terminating of a *workflow*, and verifying of a status of a *workflow* in execution.

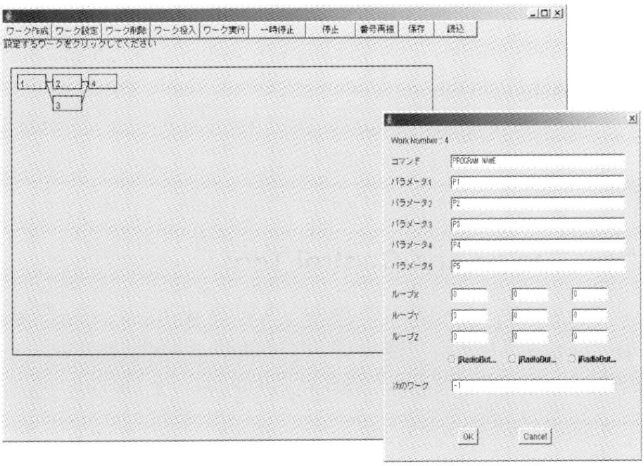

Figure 1. Worflow design and Control tool.

Here, we show the effectiveness our tool. We use "Target T0198" as the input sequence of a target structure, which was submitted at CASP6, and compare two predicted structure with and without our tool. The execution time is about 164 hours. The result is shown in Figure 2. *RMSD* in this figure represents the similarity between a predicted structure of a protein and the answer structure. The lower the score, the more similar the structures are.

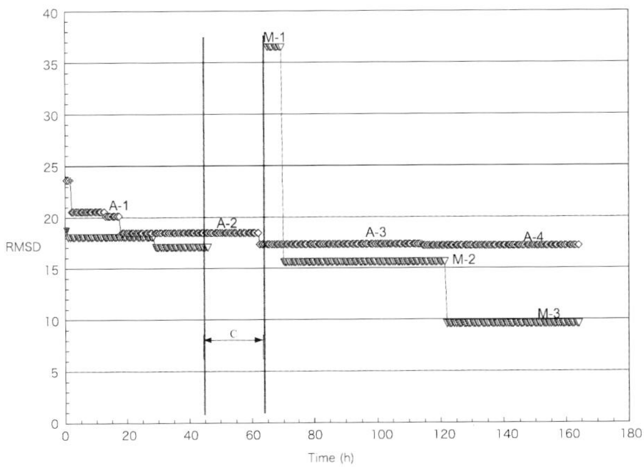

Figure 2. Simulation results.

In this figure, A-1, A-2, A-3, and A-4 are the predicted structures without our tool, and B-1, B-2, B-3, and B-4 are the ones with our tool. In the period of "C", a user modified the workflow through our tool. Thus, using our tool can improve the result of the protein structure prediction.

3. BIOPFUGA FOR MULTISCALE AND MULTILEVEL SIMULATIONS ON THE GRID ARCHITECTURE

Information of the structures and functions of protein molecules and their mutual interactions that construct protein networks increases rapidly as a consequence of the structural genomics and structural proteomics projects [7]. Advanced applications of such information require the Grid technology to solve the issue of the shortage of computational power. In particular, in the multi-scale biological systems, integration of the computational methods for models at different levels is essential.

A new platform, BioPfuga (Biosimulation Platform United on Grid Architecture), has been proposed and developed [8], where individual applications at different levels are united and executed as a hybrid application. BioPfuga requires that (i) application programs are divided into a set of many pieces, each of which corresponds to a unit simulation procedure, and that (ii) data communication be made among the program components by a standard XML description. For the first request, the program modules were implemented with a Grid service, GT3.2, defined by OGSI (Open Grid Services Infrastructure). An adaptor was useful to separate the grid middleware from the parallel computation made individually at local sites. For the second issue, a new XML description named BMSML was designed for exchanging the information among program modules at different levels in three forms: text, hexadecimal, and Base64 forms. In particular, the data size with the Base64 form amounts only to 1.3 times of that with the binary form. The schema and API tools for the XML description are available from our web site [9].

As an example of the *BioPfuga* application, we combined the quantum mechanical (QM) simulations with the program *AMOSS* for Hartree-Fock Molecular Orbital calculation (HF) developed by the NEC quantum chemistry group [10], and *GSO-X* for generalized spin density function theory (DFT) developed by Prof. Kizashi Yamaguchi [11], with the molecular mechanics (MM) simulations, the program *prestoX* [12]. We first divided these three large programs into a set of component programs. Then, as shown in Figure 3, a hybrid calculation was performed using the PC clusters located in several separated laboratories with their firewalls, and the MD part was also driven on the special-purpose computer for MD simulations, MDGrape2 [13].

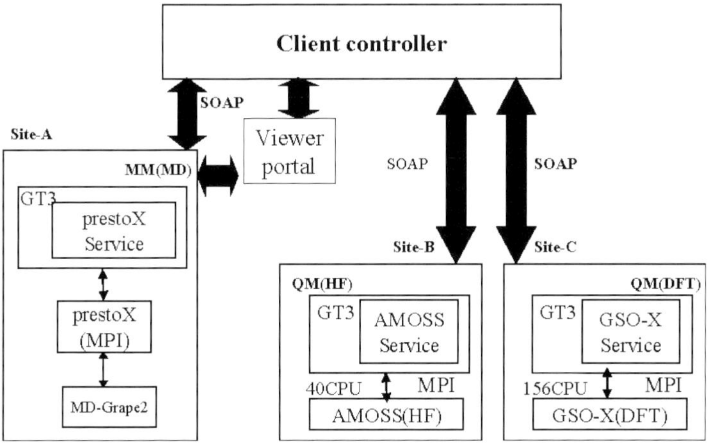

Figure 3. *BioPfuga* application system for hybrid-QM(HF)/QM(DFT)/MM method.

4. BIODATABASE FEDERATION

In this section, we present an information integration system from a number of databases for life sciences. The number of molecular biology databases has rapidly increased and now has reached more than 700 databases [14]. To solve this problem, we have devised a method for integrating several databases by connecting them all together using metadata. This connected network of databases makes use of the grid technology for delivering high performance searches of the databases. In order to bridge information among the databases, we designed three types of metadata; namely, protein, compound, and interaction metadata (see Figure 4). The interaction metadata does not describe interactions between each protein and compound individually, but denotes relationships of interactions between a set of proteins having the similar function (a protein family) and a set of compounds having an activity to the protein family. The protein and the compound metadata are used for grouping proteins and compounds according to their functional and activity classification, respectively.

The system we developed has been built giving special focus on searching for new protein-compound interactions that is an important analysis for drug discovery [15]. Our approach to this search can briefly be described as: (1) searching for a target protein related to some disease, (2) searching for homologous proteins of the target using BLAST, (3) searching for compounds interacting with the selected proteins, and (4) searching for compounds that are similar in structure to the compounds.

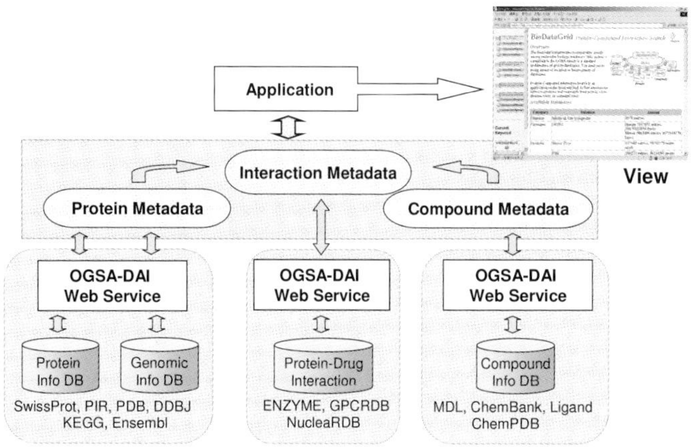

Figure 4. An information integration system for analyzing protein-compound interactions.

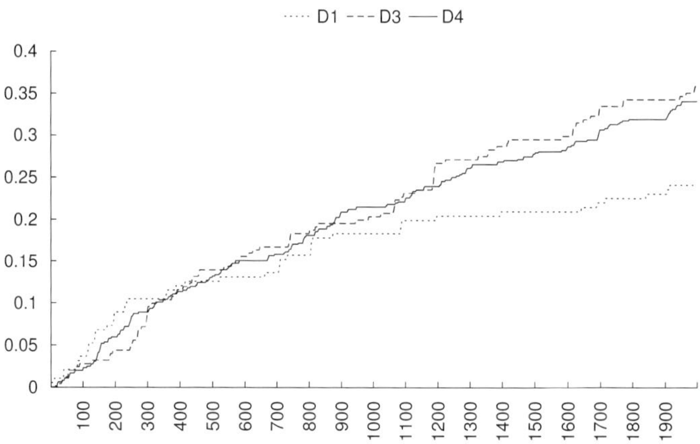

Figure 5.

To evaluate our system, we examined the analysis steps described by using this system. The four types of compounds binding to the dopamine D1, D2, D3 and D4 receptors were used. We examined a hypothesis whether two sets of compounds exhibit structural similarity if the receptors binding to the two sets are homologous. Figure 5 shows the similarity rankings that are calculated for the reference set of compounds for the D2 receptor. According to the protein homology analysis by the BLAST program, the D2 and D3 receptors are the most similar (BLAST E-value

3E-98), secondly those and the D4 receptor are similar (E-value 7E-26 between D2 and D4), and the D1 receptor is less similar to the others (E-value 1E-21 between D2 and D1). The above compound similarities are well correlated with the protein similarities.

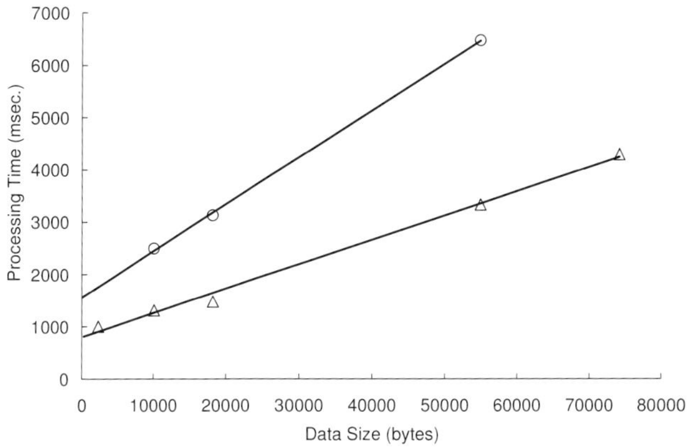

Figure 6. Processing times for the BLAST searches using DDBJ Web service (triangle, Osaka Univ.; circle, Pittsburgh).

As the performance evaluation, we measured the processing times of the BLAST searches using Web services provided by DDBJ (DNA Data Bank of Japan) (see Fig. 6). The BLAST searches were performed at Osaka University (triangle points) and at the conference center, Pittsburgh, U.S.A. (circle points) where the Super Computing conference was held.

Their results show that we have successfully extracted interaction information from the databases. We therefore conclude that our system based on grid technology is useful in the field. Further improvement in the performance of the searches is saved for future enhancements of the system.

5. SECURE GRID SYSTEM

Since the Grid emerged, it has attracted a lot of attention from researchers and scientists who are working in scientific institutions and universities. The trend is expected to continue into the future. In addition, the current situation that the Grid and web services are being integrated into a single computational platform also fascinates people working for private companies. In reality, however, a sort of technical gap between the Grid environment that they expect and the current Grid environment exists. A typical example of the research arena where such a gap exits is data security.

IPv6-based GSI-SFS, a secure file system on the Grid has been developed in hope that it satisfies the strong demands and needs on data security by bio scientists and researchers. As the name of the file system shows, the file system runs on IPv6 network as well as IPv4 network. In this section, we will describe what GSI-SFS is and how it has been used in practice.

5.1. Architecture

Figure 7 overviews the architecture of IPv6-based GSI-SFS. GSI-SFS has utilized SFS (Self-certifying File system) [16] as a building block of our secure Grid solution to satisfy the following four requirements by bio scientists and researchers.

1) Single Disk Image

Scientists and researchers seek for a consistent method for accessing data of interest regardless of data location. The method is preferred if it allows them to access data as if the data is located on a local disk.

2) Data confidentiality

Bio-related data requires confidentiality in nature. In particular, the recent situation where the application of the Grid technologies to life scientific research field, featured by the key words such as tailor-made medicine and drug discovery, is expected and demanded increasingly necessitates a technology that satisfies data confidentiality requirements.

3) Access exclusiveness

Sometimes, access information itself becomes a cause of the reveal of important information. A method for sharing data in a exclusive manner on the Grid is required.

4) On-demand remote data access

Data access should be performed in a user on-demand manner to reduce the risk of data access information leaks.

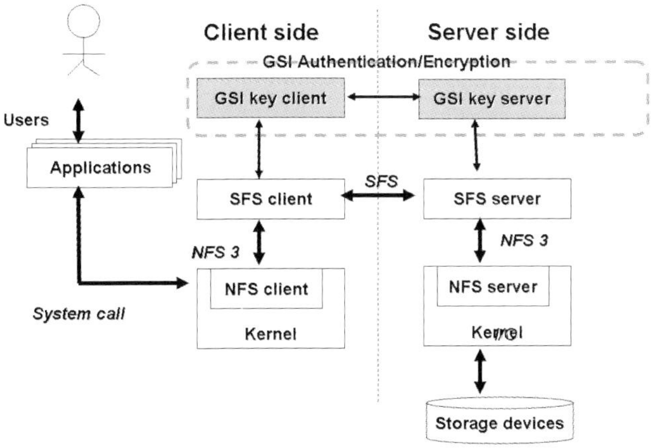

Figure 7. The overview and behavior of GSI-SFS: GSI-SFS is a secure file system for the Grid, which enhances the scalability in key management. It results in the fact that GSI-SFS has achieved the five requirements by bio scientists and researchers.

SFS is a NFS-based file system that allows scientists and researchers to access data of interest without being aware of data location. Furthermore, this file system provides a single disk image to each scientist, meaning that the access to data is performed in an exclusive way. These features are helpful and useful for secure data sharing on the Grid.

The key management problem, however, gives arise in case that we consider the use of it on the Grid where thousands of computers are connected. More specifically, a pair of RSA keys must be basically managed for each pair of sever and client in SFS, meaning that scientists needs to type a pair of user ID and password the same times as the number of the remote computers holding data they want to access.

Our solution, that is, GSI-SFS has solved this key management problem by making use of GSI (Grid Security Infrastructure). GSI provides the functionality of single sign-on and a suite of API (Application Programming Interface) to the functionality on the framework of PKIX (Public Key Infrastructure X.509) [17]. We have developed GSI authentication module so that it can mediates and replaces the original authentication process of SFS to realize the access to data in a single sign-on manner (Figure 7). As a result, GSI-SFS is enhanced in scalability and user-friendliness in comparison with original SFS.

5.2. Application Example of Secure Grid Solution

We have developed a Grid portal system that allows researchers to access text data in life science database and then perform the analytic computation using cluster systems on the Grid without being aware of data location. As of writing this, in the system, GenBank is used on a computer in China for database, and BLAST [4] and ClustalW [18] are integrated on cluster systems in Osaka University as the analytic computation. The location transparency of the system has been realized through the marriage of portal technologies and GSI-SFS. In this part, we will briefly overview how secure data sharing with location transparency is realized with GSI-SFS and Grid technologies.

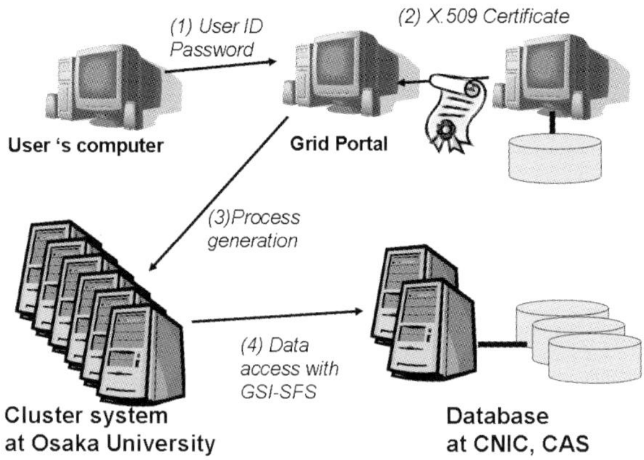

Figure 8. The overview and behavior of the Grid Portal system: This system allows scientists to access data of interest and computational resources without being aware of the location. The synergy of portal technologies and GSI-SFS has produced data transparency.

Figure 8 diagrams the overview and behavior of the Grid portal that we have developed over the network linking China and Japan. In this system, first of all, a scientist sends a pair of user ID and password to the portal system just once via intuitive web interfaces. The pair of user ID and password is used to retrieve the scientist's X.509 certificate, which is registered in advance by the scientist, from online credential repository named MyProxy [19]. The retrieved certificate is used to generate a process named grid proxy. After grid proxy is generated, the scientist is able to access data and processes on remote computers in a single sign-on manner. In the system, after the user authentication is performed on Grid portal via web interface, multiple processes of BLAST jobs are generated on the cluster systems in Osaka University. When the process needs to access data in database located in China, GSI-SFS automatically mounts the remote file system of the computer holding the database. Thus, the scientist can access data of interest without being aware of data location. Furthermore, the data is securely shared on the Grid environment. More detailed architecture of this system is described in [20].

As of writing this, this system still has future issues to solve. Examples of such issues include performance and further data security such as authorization. We would like to continue the research and development towards these issues from now on.

6. CONCLUSION

This paper summarized three examples to show how Grid technology helps the advanced life science. Many existing applications of grid technology tend to show the effectiveness of the grid only in terms of its provisioning of computational power. However, the importance of grid technology lays in its integration. It becomes a glue to combine different ICT elements. We also find that data grid technology to federate heterogeneous databases, to communicate multi-scale Simulations and to share data securely becomes more important for the advanced life science. For the advance life science, we need a concrete foundation for treating and managing large amount of and multi-scale biological information. Therefore, more integration of computing and data grid technologies is expected.

For this purpose, from 2005 onwards, we will tightly collaborate with NAREGI project (www.naregi.org), a national R&D grid middleware project stared in 2003 in terms of the development of simulation platform and data grid middleware.

ACKNOWLEDGMENTS

This work was supported by the IT program of the Ministry of Education, Culture, Science and Technology of Japan (MEXT). We appreciate the members of Biogrid project for their contributions. We also appreciate NAREGI for collaborating with us.

REFERENCES

[1] 6th Community Wide Experiment on the Critical Assessment of Techniques for Protein Structure Prediction, *http://predictioncenter.llnl.gov/casp6/Casp6.html*

[2] Fujitsuka, Y., Takada, S., Luthey-Schulten, Z.A., and Wolynes, P.G., "Optimizing physical energy functions for protein folding", *Proteins: Structure, Function and Bioinformatics*, Vol. 54, No.1, pp.88-103, 2004.

[3] Takada, S., "Protein folding simulation with solvent-induced force field: folding pathway ensemble of three-helix-bundle proteins", *Proteins: Structure, Function and Genetics*, Vol. 42, No. 1, pp. 85-98, 2001.

[4] Altschul, S.F., Madden, T.L., Schaffer, A. A., Zhang, Z., Miller, W. and Lipman, D.J., "Gapped BLAST and PSI-BLAST: a new generation of protein database search programs", *Nucleic Acids Res.*, Vol. 25, No. 17, p3389-3402, 1997.

[5] Ginalsk, K., Elofsson, Fischer, D., and Rychlewski, L., "3d-Jury: Asimple approach to improve protein structure predictions, *Bioinformatics*, Vol. 19, No. 8, pp. 1015-1018, 2003.

[6] Chikenji, G., Fujitsuka, Y., and Takada, S., "A reversible fragment assembly method for De Novo protein structure prediction, *Journal of Chemical Physics*, Vol. 119, pp. 6895-6903, 2003.

[7] Kinoshita, K., Nakamura, H., "Protein informatics towards function identification", *Curr. Opin. Struct. Biol*, vol. 13, 396-400, 2003.

[8] Nakamura, H., Date, S., Matsuda H., and Shimojo S., "A challenge towards next-generation research infrastructure for advanced life science", *Journal of New Generation Computing*, Vol. 22, No. 2, pp. 157-166, 2004.

[9] BioGrid Project, http://www.biogrid.jp/

[10] Sakuma T., Kashiwagi, H., Takada, T., and Nakamura, H., "Ab initio MO study of the chlorophyll dimmer in the photosynthetic reaction center. I. A. theoretical treatment of the electrostatic field created by the surrounding proteins", *Int. J. Quant. Chem.*, Vol. 61, pp. 137-151, 1997.

[11] Yamanaka, S., Ohsaku, Y., Yamaki, D., and Yamaguchi, K., "Generalized spin density functional study of radical reactions", *Int. J. Quant. Chem.*, Vol. 91, pp. 376-383, 2003.

[12] Fukunishi, Y., Mikami, Y., Nakamura, H., "Free energy landscapes of peptides by enhanced conformational sapling", *J. Phys. Chem. B*, Vol. 107, 13201-13210, 2003.

[13] Narumi T., Susukita, R., Ebisuzaki, T., McNiven, G., and Elmegreen, B., "Molecular dynamics machine: Special-purpose computer for molecular dynamics simulations", *Mol. Simul.*, Vol. 21, No. 5/6, pp. 401-415, 1999.

[14] M.Y. Galperin, ``The Molecular Biology Database Collection: 2005 update," *Nucleic Acids Research*, vol.33, Database issue, 2005, pp.5--24..

[15] A. Schuffenhauer, P. Floersheim, P. Acklin, and E. Jacoby, ``Similarity Metrics for Ligands Reflecting the Similarity of the Target Proteins," *Journal of Chemical Information and Computer Sciences*, vol.43, no.2, 2003, pp.391--405.

[16] Mazieres, D., Self-certifying File System, Doctoral dissertation, Massachusetts Institute of Technology, 2000.

[17] Buttler, R., Engert, D., Foster, I., Kesselman, C., Tuecke, S., Volmer, J., and Welch, V., "A national authentication infrastructure", *Computer*, Vol. 33, No. 12, pp. 60-66, 2000.

[18] Thompson, J.D., Higgins, D. G., and Gibson T. J., "CLUSTAL W: improving the sensitivity of progressive multiple sequence alignment through sequence weighting position-specific gap penalties and weight matrix choice", *Nucletic Acids Res.*, Vol. 22, No. 22, pp. 4673-4680, 1994.

[19] Novotny, J., Tuecke, S., and Welch, V., An online credential repository for the Grid: MyProxy, *Proc. of the Tenth International Symposium on High Performance Distributed Computing (HPDC-10)*, pp.104-111, 2001.

[20] Kido, Y., Date, S., Takeda, S., Hatano, S., Ma, J., Shimojo, S., and Matsuda, H., "Architecture of a Grid-enabled research platform with location-transparency for Bioinformatics", *Genome Informatics*, Vol. 15, No. 2, pp. 3-12, 2004.

A FRAMEWORK FOR BIOLOGICAL ANALYSIS ON THE GRID

TOSHIYUKI OKUMURA, SUSUMU DATE, YOICHI TAKENAKA, HIDEO MATSUDA

Graduate School of Information Science and Technology, Osaka University, 1-3 Machikaneyama, Toyonaka, Osaka 560-8531, Japan

Email: { okumura55 , sdate, takenaka, matsuda }@ist.osaka-u.ac.jp

With the rapid progress of the human genome project and related analyses, a huge amount of sequence data has been generated and a substantial number of methods has been proposed for predicting the potential functions based on sequence homology, functional patterns (motifs), domain information and so forth. It is often the case that actual processes of these biological analyses are not straightforward but rather complicated. In order to solve this problem, we propose a framework to virtualize and integrate various biological resources such as programs, databases and experimental data on the Grid environment. We show how our architecture makes it possible to improve the complicated process of biological analyses.

1. INTRODUCTION

Since the human genome project[1] and related analyses were completed, a vast amount of data has been and is being generated. The substantial portion of such data is sequence data and the functional analysis of sequences is an important issue for molecular biologists. Dozens of methods have been proposed for analysis and predicting the potential functions from sequences. The predictive performance of biological analysis programs are very varied and depend on the type of the function to be identified, such as transcription, regulation, enzyme, and signal transduction. It is often the case that actual processes of biological analyses are not straightforward but rather complicated. In order to solve this problem, the integration of diverse analysis programs, experimental data and other additional information is required.

Grid technologies[2] provide us the virtualization mechanisms of various resources such as computing, data, network and organizations. These technologies play an important role for the virtualization of scientific resources and enhance the efficiency of advanced research. We propose a framework to virtualize and integrate various resources for biological analyses on the Grid environment.

The rest of this paper is organized as follows. In Section 2, we describe the problems of biological sequence analyses by use of our case study and derive the motivation of our research. Then we present a framework for the integration of sequence analyses on the Grid environment in Section 3. Section 4 discusses our architecture and future issues. We conclude our work in Section 5.

2. MOTIVATION

We proposed and performed a systematic method for detection of conserved sequence region to discover new, biological relevant motifs from a set of 21,050 conceptually translated mouse cDNA (RIKEN FANTOM1) sequences.[3,4] Figure 1 shows the main procedure flow of motif exploration for the RIKEN FANTOM1 project. In this analysis, we used several methods to explore new motifs as shown below.

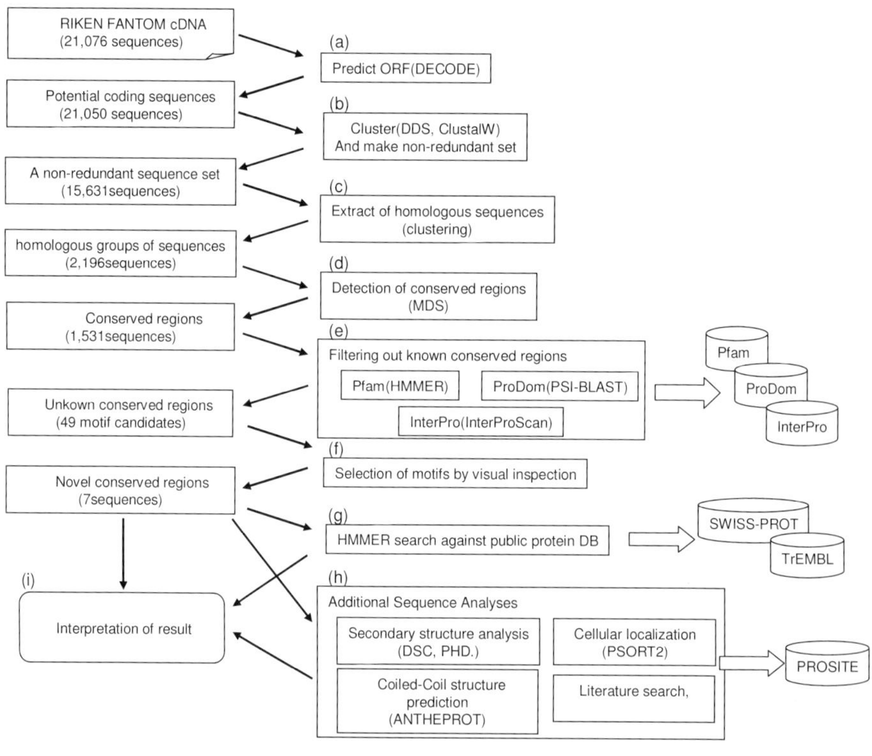

Figure 1. The process flow of motif exploration.

(a) Potential Protein Sequences
The FANTOM cDNA collection contains 21,076 clones. We used SEG[5] and DECODER to predict protein coding sequences (21,050).

(b) Preparation of a non-redundant sequence set
We clustered these putative protein sequences using DDS[6] and ClustalW.[7] We obtained 15,631 non-redundant sequences by selecting the longest sequence as representative of the cluster.

(c) Extract of homologous sequences
We used a linkage-clustering method with all-to-all sequence comparison[8] to extract 2,196 homologous groups of sequences from a non-reduntdant set of RIKEN mouse cDNA sequences.

(d) Detection of conserved regions (motif candidates)
1,531 homologous regions in groups were detected with a maximum-density subgraph (MDS) algorithm[9] which is a graph-theoretic method.

(e) Filtering out known conserved regions
The detected blocks were screened for known conserved regions detected by HMMER in Pfam,[10]PSI-BLAST[11] in ProDom,[12] and InterPro Scan in InterPro databases.[13] Blocks that overlapped with at least one residue of known motifs or domains were discarded. We obtained 47 motif candidates through these processes.

(f) Select motifs by visual inspection

(g) HMMER search against public protein databases
We constructed HMMs from the conserved blocks and searched against SWISS-PROT/TrEMBL databases[14] with HMMER[15] to expand the number of motif members.

(h) Additional sequence analyses
We performed additional sequence analyses as follows to reveal characteristics of novel motifs.
- Secondary structure analysis by DSC[16] and PHD[17]
- Cellular localization prediction by PSORT2[18]
- Coiled-Coil structure prediction by ANTHEPROT[19]
- Literature search

(i) Interpretation of result
We considered all the various factors together and interpreted them.

As illustrated in Figure 1, the detection of motif candidates is relatively easy even on genome scale. On the other hand, the exploration of new motifs for potential biological functions is a time-consuming process that needs to manage a large variety of coordination processes and experimental validation. This is our next issue to be solved.

3. SYSTEM ARCHITECTURE

We propose a simple framework to manage these various programs and experimental data efficiently on the Grid. Grid technology is one of the most promising technologies that enable us to efficiently integrate heterogeneous resources.

3.1. Virtualization of Service Components

At first, various sorts of resources including analysis programs and databases are encapsulated and virtualized as Service Components on the Grid infrastructure (Figure 2). Each service component complies with the rules of the Grid service. The virtualization enables us to access various resources in the same manner and it increases effectiveness of biological analysis process to considerable extent.

Another advantage of the service component is that some sort of the Set Operation is available (conceptually). We are able to obtain subset from the output of service components with elementary operations such as union, intersection, difference, and complement. Even though only limited operations are available, it has a potential for changing the way of biological data analysis.

As described in this section, the virtualization makes some improvements for our first issue. But it is not enough because there still remain a large variety of processes to be coordinated. That is our second issue.

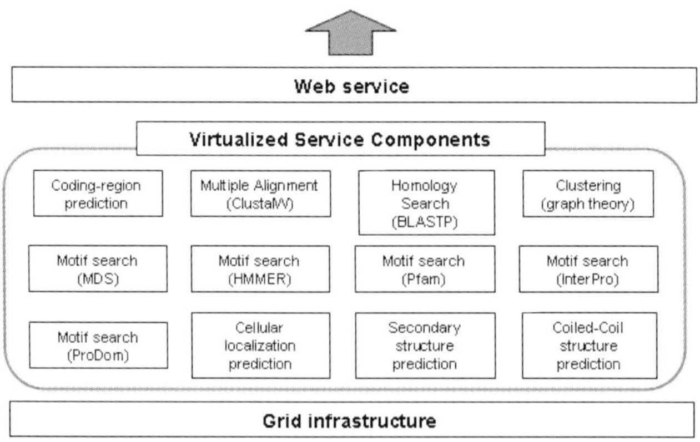

Figure 2. Virtualization of service components.

3.2. Service Metadata and Service Mediator

One of the key concepts we propose in this paper is Service Metadata and Service Mediator (Figure 3). Service Metadata is designed for the federation of virtualized service components. Service Metadata describes the profile information of each component and the relationship information between individual service components. The profile information includes service type, functional description, input/output data format, parameters and other essential information for the use of each service. The relationship information enables us to combine several components and generate another virtual component. Virtualized resources such as programs, databases, and experimental data are coordinated and rationalized with Service Metadata.

And then, Service Mediator plays an important role to integrate distributed service components with use of Service Metadata and manage the state of workflow with State Information. As shown in Figure 3, Service Mediator receives requests from applications and mainly provides following services.

(a) return the information of available services
(b) call for appropriate service components and return the result information
(c) combine several service components and perform them as a virtual service
(d) update the information of available services
(e) save and update State Information

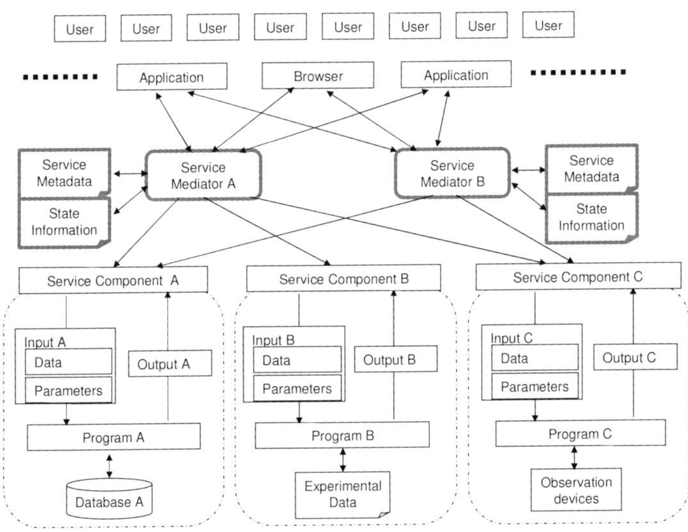

Figure 3. Outline of Service Metadata and Service Mediator.

3.3. Workflow Management with State Information

The workflow management is another important feature of Service Mediator (Figure 4). Service Mediator makes records of the workflow history based on each user's attribute (Figure 4 (A), (B)) and stores them in State Information (Figure 3). The State Transition Matrix (STM) information is generated from the workflow history (Figure 4 (C). This STM is generalized information without any privacy data of individual person. In some aspects, this STM represents some kind of context in biological analysis procedures. By using this STM, Service Mediator is able to generate a guidance of workflow which indicates next procedure from current service component (Figure 4 (D), (E)). In addition, this STM enables us to generate a representative path dynamically which is commonly used in a community. Because each STM reflects actions of certain community, we can use them as sort of knowledge.

Although the STM is a well known algorithm and we have to improve and evolve this to more sophisticated one, this function is convenient for biologists.

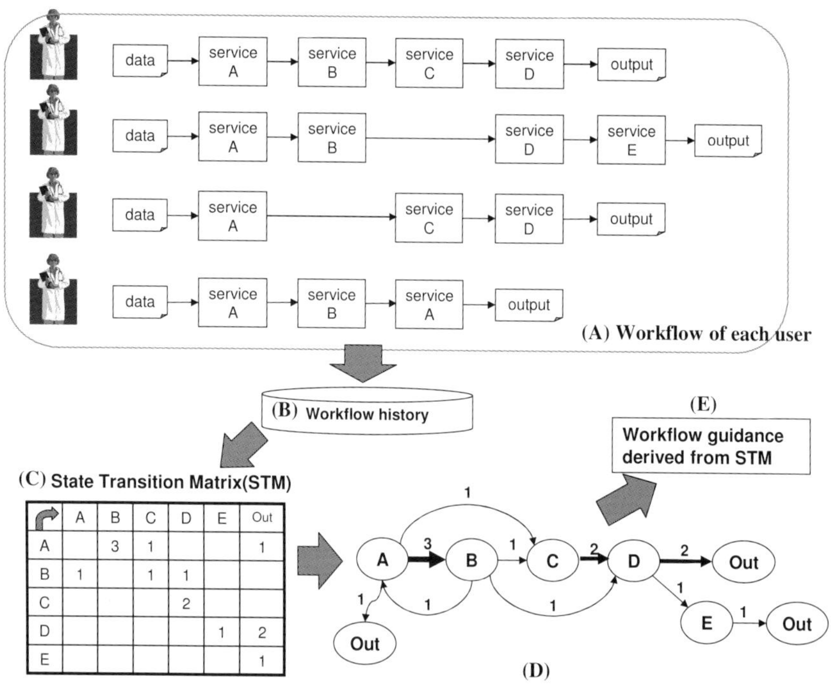

Figure 4. Workflow management of Service Mediator.

4. DISCUSSION

We described a framework which enables us to reduce the complexity of biological analyses. In order to evolve our framework, we need further discussion about characteristic of biological data. The common characteristic of biological data is inclusion of a large number of attributes and the optimal analysis method or procedure depends on data and circumstances. This is the remaining issue we have to address in the next stage.

5. CONCLUSIONS AND FUTURE WORK

In this paper, we proposed a concept and framework to integrate biological resources and described an advantage of our architecture. This scheme is applicable not only for the sequence analysis but for various biological analyses and other scientific research fields.

However, the implementation is not completed yet. We are going to apply this framework to our real issue described in Section 2. We are preparing and designing this software under Globus Toolkit 4.0 and WSRF.[20]

Our goal is to improve the precision of biological analysis and acquire deep knowledge through virtualization and integration of the relevant resources and information. We believe this kind of approach must be a key solution for biologists who are constrained to manage vast amount of data in post-genome era.

ACKNOWLEDGMENTS

We are grateful to Yoshiyuki Kido, Yoshitada Fukuoka and Takuya Sato for their advice and support.

REFERENCES

1. International Human Genome Sequencing Consortium. Finishing the euchromatic sequence of the human genome. *Nature* 2004; **431**: 931–945.
2. Global Grid Forum. http://www.gridforum.org/
3. RIKEN Genome Exploration Research Group Phase II Team and FANTOM Consortium. Functional annotation of a full-length mouse cDNA collection. *Nature* 2001; **409:** 685–690.
4. Kawaji H, Schönbach C, Matsuo Y, Kawai J, Okazaki Y, Hayashizaki Y, Matsuda H. Exploration of novel motifs derived from mouse cDNA sequences. *Genome Research* 2002; **12**: 368–378.

5. Wootton JC, Federhen S. Analysis of compositionally biased regions in sequence databases. *Methods in enzymology* 1996; **266**: 554–571.
6. Huang X, Adams MD, Zhou H, Kerlavage AR. A tool for analyzing and annotating genomic sequences. *Genomics* 1997; **46(1)**: 37–45.
7. Thompson JD, Higgins DG, Gibson TJ. CLUSTALW: improving the sensitivity of progressive multiple sequence alignment through sequence weighting, position-specific gap penalties and weight matrix choice. *Nucleic Acids Research* 1994; **22**: 4673–4680.
8. Kawaji H, Takenaka Y, Matsuda H. Graph-based clustering for finding distant relationships in a large set of protein sequences. *Bioinformatics* 2004; **20(2)**: 243–252.
9. Matsuda H. Detection of conserved domains in protein sequences using a maximum-density subgraph algorithm. *IEICE Trans. Fundamentals Electron. Commun. Comput. Sci.* 2000; **E83-A**: 713 721.
10. Bateman A, Birney E, Cerruti L, Durbin R, Etwiller L, Eddy SR, Griffiths-Jones S, Howe KL, Marshall M, Sonnhammer EL. The Pfam protein families database. *Nucleic Acids Research* 2002; **30(1)**: 276–280.
11. Altschul SF, Madden TL, Schaffer AA, Zhang J, Miller W and Lipman DJ. Gapped BLAST and PSI-BLAST: a new generation of protein database search programs. *Nucleic Acids Research* 1997; **25(17)**: 3389–3402.
12. Servant F, Bru C, Carrere S, Courcelle E, Gouzy J, Peyruc D, Kahn D. ProDom: automated clustering of homologous domains. *Briefings in Bioinformatics* 2002; **3(3)**: 246–251.
13. Zdobnov EM, Apweiler R. InterProScan — an integration platform for the signature-recognition methods in InterPro. *Bioinformatics* 2001; **17(9)**: 847–848.
14. O'Donovan C, Martin MJ, Gattiker A, Gasteiger E, Bairoch A and Apweiler R. High-quality protein knowledge resource: SWISS-PROT and TrEMBL. *Briefings in bioinformatics* 2002; **3(3)**: 275–284.
15. Eddy SR. HMMER: Profile hidden Markov models for biological sequence analysis. http://hmmer.wustl.edu/, 2001;
16. King RD, Saqi M, Sayle R, Sternberg MJ. DSC: public domain protein secondary structure predication. *Computer applications in the biosciences* 1997; **13(4)** 473–474.
17. Rost B. PHD: predicting one-dimensional protein structure by profile-based neural networks. *Methods in enzymology* 1996; **266**: 525–539.
18. Nakai K, Horton P. PSORT: a program for detecting sorting signals in proteins and predicting their subcellular localization, *Trends in biochemical science* 1999 **24(1)** 34–36.
19. Deleage G, Combet C, Blanchet C, Geourjon, C. ANTHEPROT: an integrated protein sequence analysis software with client/server capabilities, *Computers in biology and medicine* 2001; **31(4):** 259–267.
20. The Globus Alliance. http://www.globus.org/

AN ARCHITECTURAL DESIGN OF OPEN GENOME SERVICES

RYO UMETSU, SHINGO OHKI, AKINOBU FUKUZAKI, AKIHIKO KONAGAYA

Advanced Genome Information Technology Group, RIKEN, 1-7-22, Suehiro-cho, Tsurumi, Yokohama, Kanagawa, Japan
Email: {u-ryo, ohki, akki, konagaya}@gsc.riken.jp

DAISUKE SHINBARA, MASATAKA SAITO, KENTARO WATANABE, TETSUJI KITAGAWA, TEPPEI HOSHINO, AKIHIKO KONAGAYA

Department of Computer Science, Tokyo Institute of Science and Technology, 2-12-1, Ookayama, Meguro, Tokyo, Japan
Email: {shinbara, saitou, kwatanabe, tetsu, teddy, konagaya}@bio.cs.titech.ac.jp

This paper proposes a new research platform for genome science, named Open Genome Service, based on Web Services and a grid system with workflows. Web Services with a grid system as back-end servers have greater potential to solve issues on scalability or accessibility over networks than Web Applications or local bioinformatics environments. Workflows and a filter mechanism support this framework. We have implemented an initial prototype system for genome annotation and showed its prospect.

1. INTRODUCTION

Data driven biology has produced enormous amounts of genomic information including genome, transcriptome, proteome and phenome, to name a few. More than hundreds of biological databases and bioinformatics tools are now available on the Internet. It is highly desired to integrate all of them and extract genetic knowledge valuable for biomedical applications such as drug design and personalized medicine. To promote this, a paradigm shift is necessary in programming style from conventional stand-alone programming to loosely-coupled network programming that takes advantages of database and application services available on the Internet.

Some advanced research institutes have already started bioinformatics web services with advanced web technologies, such as SOAP, WSDL and UDDI, as seen in EBI[8], KEGG[16] and DDBJ[23], and others. These web services enable us to fetch genetic data from databases and execute BLAST, FASTA and other bioinformatics tools without installing them on a local computer. This indicates the advent of "global bioinformatics environment" in which all biological databases and bioinformatics tools are available as network services on the Internet in the near future.

Grid is another important information technology that promotes bioinformatics[2]. With increasing database sizes, bioinformatics applications such as BLAST and FASTA, require parallel computing or cluster computing in order to accelerate search performances. Systems biology also requires high performance computing for mathematical simulation and kinetic parameter estimation.[12, 17] Grid technology brings us powerful solutions for this purpose by means of exploiting parallel and distributed processing with secure protocols, e.g., GridBlast[11], OBIEnv[18], GOGA (Grid-Oriented GA)[13].

This paper proposes a new research platform for genome science, named Open Genome Service, based on Web Services and a grid system. The goal and motivation of this study is very similar to Open Grid Service Architecture (OGSA) [19] but we do not adopt OGSA-DAI in this study in order to focus on practical engineering issues to deal with genome applications on Web Services rather than pursuing grid service technology. We strongly believe that our Web Service architecture is versatile enough to provide a grid service interface whenever required.

In Section 2 of this paper, we describe our motivation and design philosophy to promote Open Genome Service. Section 3 depicts our approaches to workflow and orchestration. Then we discuss an architectural design of our prototype system in Section 4. Section 5 demonstrates our early results on automatic genome annotation. Lastly, Section 6 gives our conclusion on Open Genome Service and future works.

2. DESIGN PHILOSOPHY ON OPEN GENOME SERVICE

Two major approaches have been known to use biological databases and bioinformatics tools.

1. A local bioinformatics environment constructed on UNIX servers or PC clusters
2. Web Applications on the Internet

The first approach has great advantages in data access performance especially for high throughput and large scale data processing (e.g., for genome sequence annotation). However, it needs skillful operators to integrate and to maintain biological databases and tools, as well as high performance computers and large secondary storage facilities. It often happens that the databases and tools are so tightly-coupled that we have to port not only a tool but also its bioinformatics environment. This is one reason why we have to maintain similar copies of biological databases for bioinformatics tools.

On the other hand, the second approach has an advantage in its openness. The web application enables us to use databases and bioinformatics tools from anywhere over the Internet without worrying about computational resources nor maintenance costs. However, human interactive processing on web browsers often becomes a bottleneck especially when dealing with large data amounts like genome sequences.

Web Services have a great potential to solve these issues and to combine the best of both approaches. Web Services enable remote access of databases and analysis tools without human interactive processing. The Web Services enable us to develop bioinformatics tools independent of a local bioinformatics environment. In other words, the Web Services enable us to construct a bioinformatics environment which has the following capabilities:

- Loose coupling of databases and applications at runtime (late binding)
- Portable bioinformatics application development over the network (location-free)

However, in order to apply Web Services for genome sciences, the following issues must be addressed.

- Handling of very large data (data scalability).
- Handling of massive processes (computation scalability).
- Mechanism to modify computational results (bioinformatics knowledge).

In order to solve the above issues, we adopted a grid system as back-end servers for Web Services. Although the integration of Grid and Web Services is one of hot topics in grid research, many issues remain to be solved from a view point of genome services: asynchronous processing of long time jobs, handling of large computational result sets, load balancing of servers, to name a few. We will show our solution in Section 4.

Another important research topic we are now pursuing is the design of workflow and orchestration to describe the "business logic" of genome services. Genome services are not a simple mixture of bioinformatics applications. Special experiences and know-how are needed to select biologically meaningful results from computational results and interpret them. For example, in case of a homology search, the lowest expectation-value does not always show the best search result.

Skillful bioinformaticians often select the best candidates by eliminating artifacts in the search results with their biological knowledge and experiences. Workflow or business logic of bioinformatics should provide such facilities that deal with the biological knowledge and experiences.

Although it is too early to draw any conclusion at this stage, our solution for this issue is to provide an object-oriented framework that transforms computational results with user's knowledge and experiences. Design patterns will play an essential role to represent the knowledge of bioinformatics experiments.

Finally, although we do not discuss the details in this paper, security is one of the most important research topics for us to promote genome services, especially when dealing with personal information. Total solution is required from terminals to servers as well as protocols for authentication and authorization. Enterprise Java Beans (EJB) may give us a partial solution for this purpose. However, security technology for genome services is one of the research challenges we must pursue from a long-term point of view.

3. WORKFLOW AND ORCHESTRATION

Web Service design strongly depends on workflow and orchestration of bioinformatics tools for computational experiments. As the basis of our system architecture design, the following policies were adopted:

1. Provide interfaces to make use of existing bioinformatics applications as workflow components (reusability)
2. Separate the application logic from grid implementation (encapsulation)
3. Provide various levels of Web Services for users (granularity)
4. Adopt existing frameworks, such as J2EE, and design patterns (stability)

1. There is a plethora of standard applications in the bioinformatics field today. There is a little point to rebuild all of them from scratch. In our system, Web Service wrappers are provided to make use of well-known bioinformatics applications tools.

2. In designing a workflow framework, we encapsulate all grid interfaces in Web Services so that workflow designers can concentrate on their efforts for biological issues without worrying about implementation issues related to parallel and distributed processing. This may cause some overhead in performance but we strongly believe that minimum of interfaces is a secret of longevity for software systems.

3. Granularity of Web Services. If the granularity is too large, the flexibility for users becomes low. On the other hand, if it is too small, users will have unnecessary

problems with linking web services. To provide a class hierarchy of Web Services and workflows appears to be one practical solution to solve this well-known problem.

4. The adoption of existing frameworks and intensive use of design patterns. Distributed processing is one of the most difficult systems to develop in terms of performance, stability and security. Existing frameworks enable us to develop sophisticated web services at reasonable costs. Design patterns are also useful to ensure modularity and reusability of classes as well as to minimize total development costs.

Other important choices in workflow framework design are the following.

Data formats among Web Services: Common XML format is one attractive solution for assuring interoperability, but great efforts are needed for standardization. We adopted an ad hoc solution in the current implementation by providing appropriate filters which translate some computational results into the input forms of the next Web Services. Fundamental solutions should be studied from a long-term research point of view.

Synchronous vs. asynchronous Web Services: In bioinformatics, some applications may require long time execution and some applications may produce thousands of subcommands from one upper level command execution. In order to deal with such cases, asynchronous message handling is indispensable in bioinformatics.

Workflow description language: There are many proposals in business workflow description including BPEL (Business Process Execution Language)[7] and XPDL (XML Process Description Language)[22], to name a few. There is room for arguments that, a business process description language can also be applicable for bioinformatics workflow description or we need something original for bioinformatics.

Workflow service type: Which is better to execute a workflow program: client side or server side? Either of these implementations is possible but there are many differences in system designs; user interface, performance, security, etc.

Table 1 summarizes on-going projects related to workflow orchestration: The Taverna Project[21] has developed a powerful platform with excellent GUI for workflow compositions and result visualizations. Its workflow engine is called Freefluo, which integrates Web Services and other types of public services as such BioMOBY[5]. BPWS4J[6] and ActiveBPEL[1] are workflow engines combining web services by BPEL. Enhydra Shark[9] is a well-known workflow engine using XPDL. jBPM[14] is a Java business management engine hosted by JBoss. It uses jpdl, another original workflow description language.

Table 1. Comparison Table

Name	Server Type	Language	Workflow Editor
Taverna (Freefluo)	standalone/web application server	original	Workbench
BPWS4j	application server	BPEL	None
ActiveBPEL	web application server	BPEL	ActiveWebflow
Enhydra Shark	standalone / CORBA	XPDL	Enhydra JaWE
jBPM	standalone / J2EE server	jpdl	None

4. PROTOTYPE SYSTEM

According to the system design described above, we have developed a prototype that provides the following features.

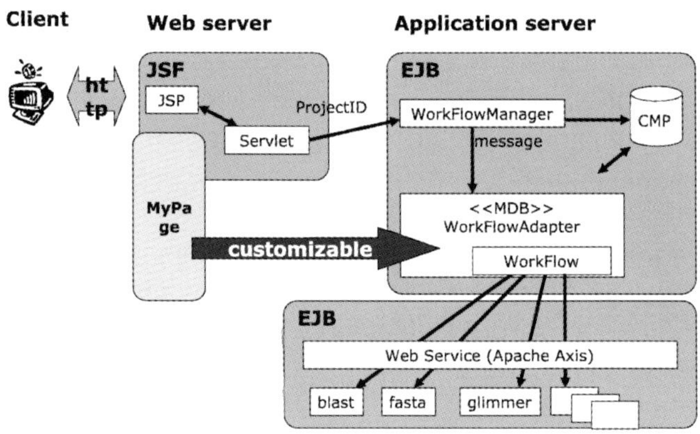

Figure 1. Designed System

4.1. Five-Layer Model

The prototype system consists of five layers from a view point of SOA (Service Oriented Architecture).[3] A five-layer model is suitable for our system design that requires to organically and loosely combines distributed components of bioinformatics services into a meaningful workflow process using Web Services. According to this five-layer model architecture, business logics can be easily distributed although they are also able to be integrated as one service. And we can also provide a web interface layer as described below.

4.2. J2EE framework (EJB)

We adopted a J2EE framework to develop a robust and scalable distributed system necessary for Genome Web Services in a short period of time. The J2EE framework provides various functionalities: object caching, clustering, load balancing, remote method invocation, transaction management, user securities (authentication and authorization), persistence on any kind of database implementations, messaging services, web container, naming and directory interfaces, JSP (Java Server Pages), and more. This makes it possible to concentrate our efforts on core system development and to reduce the total development costs considerably. We also intensively use design patterns to avoid agglutination issues which often occur in a J2EE framework when implementing program codes to the framework.

4.3. Workflow Class Library

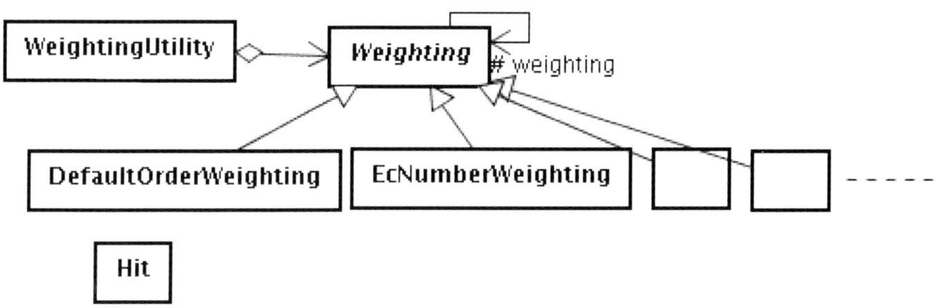

Figure 2. Filter class hierarchy

In order to demonstrate the effectiveness of Web Services in genome analysis, the prototype system provides not only Web Services for bioinformatics commands but also a workflow, in which Web Services are organically chained for open reading frame (ORF) annotation. Some filtering classes are provided for the interaction of computational results. Another filter can be easily added by expanding the `Weighting` class (as Figure 2) and composing them. With this module, users can decide how the best result should be picked up. Enhancement of the workflow design tool and expansion of the workflow and filter class library are the most urgent issues we are now focusing on.

4.4. Back-End Grid System

The current prototype system provides a PC cluster as a back-end of the Web Services. Based on our experience on grid research, expansion of a back-end PC cluster to a back-end grid system is almost straight forward if an appropriate job scheduler and a global file server are provided. However, care must be taken so that overhead for distributed parallel computing does not affect the total system performance. The overhead would be negligible when the Web Services have to deal with plenty of jobs for multiple users.

4.5. Web Interface

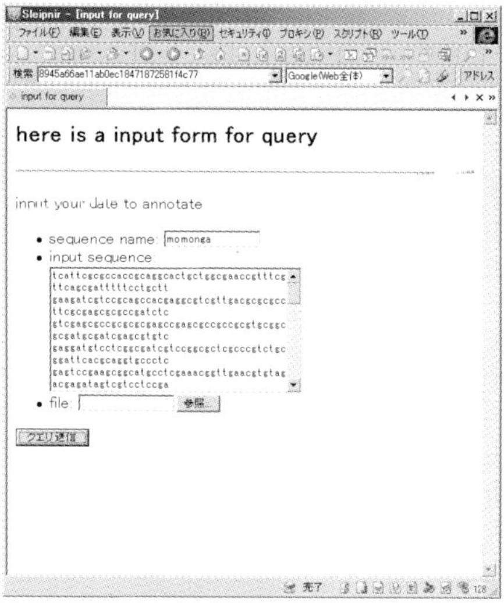

Figure 3. Input Form

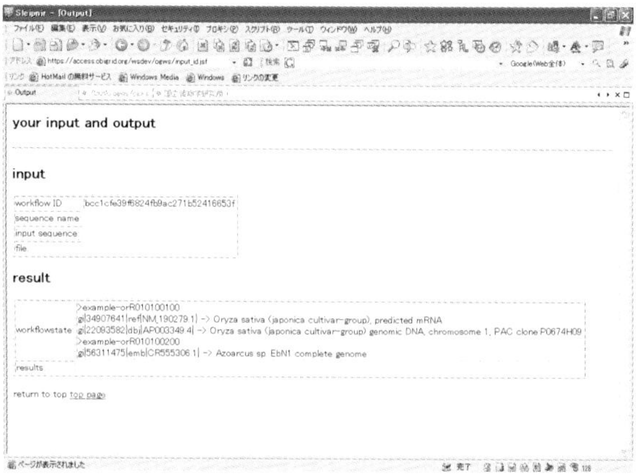

Figure 4. Output Form

A web interface (web application interface) is also available in our prototype system so that a user can access a workflow through his/her web browser. This interface is written in JSF, a new standard of presentation layer in Java.[15] User enters their query sequence into a web browser interface, get a workflow id, and access the result of annotations by the id. A JSF backing bean class sends a message to dispatch a workflow to JMS (Java Messaging Service) and returns a workflow id to an interface. JMS wakes up a MDB (Message Driven Bean: a kind of EJB) and the MDB executes a workflow in the background. A security system will be provided soon to guard their own data and a customizable page framework which consists of pieces (Portlet[20]) of web service interfaces to compose a workflow.

5. EVALUATION

Figure 5 shows the workflow diagram and the execution performance respectively for genome annotation in our prototype system, respectively. The web page takes two inputs, a nucleotide sequence and an ORF base name for glimmer command and returns a workflow ID which is used for monitoring job status and obtaining computational results such as ORF-names and their definitions when the workflow is done.

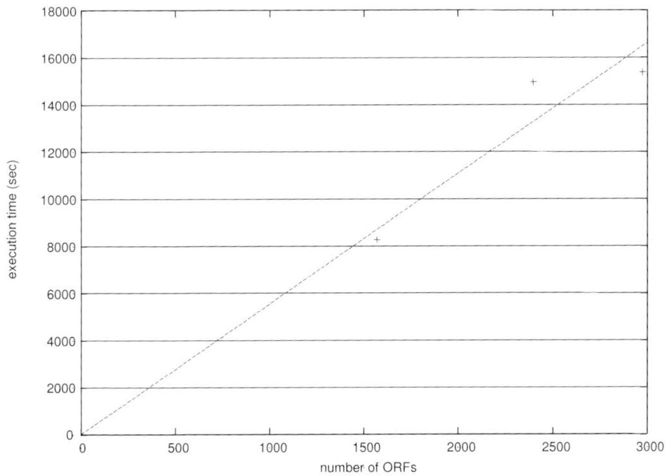

Figure 5. Execution Performance

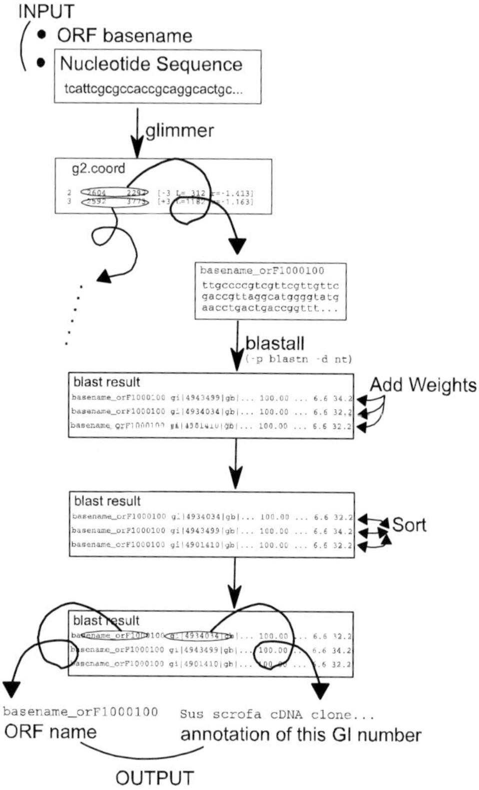

Figure 6. Workflow Diagram

The workflow is described below.

1. Take two arguments; a nucleotide sequence and a base name of ORFs.
2. Invoke a glimmer web service which executes run-glimmer2 on the query sequence.
3. Parse the glimmer result, generate an ORF name on each resultant ORF, and extract ORF FASTA format sequences from original nucleotide sequence.
4. Invoke a blastall web service with a "`-p blastn -d nt`" option and single-FASTA format sequence on each result.
5. Apply a "filter" which sorts the results of blastn by certain criteria (e-value, existence of EC number, to name a few).
6. Pick up the first item of the sorted result sets and combine it with an ORF number as an initial computational annotation.

7. Store the ORF number and its annotation to Container Managed Persistence (CMP).
8. Wait until user enters a workflow id.

6. CONCLUSION

In this paper, we described the design of Genome Service as genome sequence annotation as an example. A J2EE framework plays an essential role in our prototype system to develop a distributed processing system necessary for Genome Service. Workflow and orchestration of bioinformatics tools demonstrate a great potential of a global bioinformatics environment constructed on the Internet.

Future tasks toward the above goal are:

- Enabling to process large scale jobs with back-end grid
- Realizing a flexible and customizable workflow platform
- Strengthening security for web services
- Improving the performance of asynchronous Web Services

REFERENCES

1. ActiveBPEL Engine --- Open Source BPEL Server. `http://activebpel.org/`
2. Akihiko Konagaya and Kenji Sato (Eds), Grid Computing in Life Science. *Lecture Notes in Bioinformatics* 2005; vol.3370.
3. Ali Arsanjani. Service-oriented modeling and architecture. `http://www-106.ibm.com/developerworks/library/ws-soa-design1/`
4. Applab Homepage. `http://industry.ebi.ac.uk/~senger/applab/`
5. BioMOBY.org Main page. `http://www.biomoby.org/`
6. BPWS4J. `http://www.alphaworks.ibm.com/tech/bpws4j`
7. Business Process Execution Language for Web Services Version 1.1. `http://www-128.ibm.com/developerworks/library/ws-bpel/`
8. EBI Tools. `http://www.ebi.ac.uk/Tools/webservices/`
9. Enhydra Shark. `http://shark.objectweb.org/`
10. Freefluo Workflow Enactor. `http://freefluo.sourceforge.net/`
11. GridBlast, `https://access.obigrid.org/BLAST/`
12. Hatakeyama M, Kimura S, Naka T, Kawasaki T, Yumoto N, Ichikawa M, Kim JH, Saito K, Saeki M, Shirouzu M, Yokoyama S, Konagaya A. A computational model on the modulation of mitogen-activated protein kinase

(MAPK) and Akt pathways in heregulin-induced ErbB signaling. *Biochemical Journal* 2003; **373**: 451-463.
13. Hiroaki Imade, Naoaki Mizuguchi, Isao Ono, Norihiko Ono, Masahiro Okamoto, "Gridifying" A Parallel NSS-EA using The Improved GOGA Framework And Its Performance Evaluation on OBI Grid. *1st Int'l Workshop on Life Science Grid (LSGRID2004)* 2004; 159-160.
14. Java Business Process Management. `http://jbpm.org/`
15. JavaServer Faces. `http://java.sun.com/j2ee/javaserverfaces/`
16. KEGG API. `http://soap.genome.jp/`
17. Kimura S, Ide K, Kashihara A, Kano M, Hatakeyama M, Masui R, Nakagawa N, Yokoyama S, Kuramitsu S, Konagaya A, Inference of S-system models of genetic networks using a cooperative co-evolutionary algorithm. *Bioinformatics* 2004; vol.20, no.10: 1646-1648.
18. OBIEnv, `https://access.obigrid.org/jaist/obienv/`
19. OGSA. `http://www.globus.org/research/papers/ogsa.pdf`
20. The Java Community Process(SM) Program - JSRs: Java Specification Requests - detail JSR#168. `http://www.jcp.org/en/jsr/detail?id=168`
21. The Taverna Project Website. `http://taverna.sourceforge.net/`
22. Workflow Process Definition Interface - XML Process Definition Language 1.0 Final Draft.
`http://www.wfmc.org/standards/docs/TC-1025_10_xpdl_102502.pdf`
23. XML Central of DDBJ. `http://xml.nig.ac.jp/`

MAXIMIZING COMPUTATIONAL CAPACITY OF COMPUTATIONAL BIOCHEMISTRY APPLICATIONS: THE NUTS AND BOLTS

TERRY MORELAND AND CHIH JENG KENNETH TAN[†]

OptimaNumerics Ltd
Belfast, United Kingdom
Email: {tmoreland|cjtan}@OptimaNumerics.com

The canonical form of many large-scale biochemistry applications are often linear algebra problems. The performance of matrix solvers is often critical in determining the performance of the application programs. This paper investigates the performance of common linear algebra routines on the current architectures of interest to supercomputing users, namely the Intel Xeon EM64T, Intel Itanium II and IBM PowerPC970, with examples from OptimaNumerics Libraries. The necessity to consider the performance of underlying nuts and bolts components of a biochemistry application prior to committing investments to deploy the application into a Grid computing environment is highlighted. Performance issues and myths are also discussed in this paper.

1. INTRODUCTION

Over the last few years, there has been an increase in the number of life sciences applications taking advantage of what is generally classified as "Grid computing". Within the life sciences domain, one of the sub-domains is computational biochemistry. While Grid computing enables a practitioner to solve much larger problems, it is still crucial to ensure that the application is running optimally on the Grid. As such, we need to look under the hood to investigate the nuts and bolts needed to attain application optimality.

Looking under the hood, we find that the bulk of computational biochemistry applications, similar to other scientific and technical computing applications, can be reduced to three classes of "basic" problems: linear algebra, Fourier transform, and randomized simulation. Of these three classes, linear algebra is often the dominant class of problems solved in many computational biochemistry applications. We find

[†] Also at: School of Computer Science, The Queen's University of Belfast, Belfast, UK.

these situations in software packages such as GAMESS, Wien2k, PWscf, ADF, SIESTA, TURBOMOLE, among others.

End applications ranging from drug design to structural simulations, involve linear algebra functionalities such as linear solvers, eigen solvers and diagonalizers at their core. We need to ensure that the code performing these operations are at their top performance, to have the application run efficiently.

Over the years, Linear Algebra Package (LAPACK),[6] Scalable Linear Algebra Package (ScaLAPACK)[8], has arisen as de facto standards for handling common linear algebra operations in a portable manner.

Today, LAPACK and ScaLAPACK routines are widely accepted and used as standard linear algebra routines. LAPACK and ScaLAPACK implementations are available from OptimaNumerics and other high performance computing software vendors. Also, some hardware manufacturers provide versions of LAPACK and ScaLAPACK that are intended to be specifically tuned for the particular architecture, along with the lower level Basic Linear Algebra Subroutines (BLAS).[1]

Here we will investigate the performance of common linear algebra routines on the current architectures of interest to supercomputing users, namely Intel Xeon EM64T, Intel Itanium II and IBM PowerPC970. The problems and myths of performance are also shown and diffused in this paper. This is a work in progress, following on previously reported results.[9, 10]

2. COMMON OPERATIONS IN BIOCHEMISTRY APPLICATIONS

The common linear algebra operations that biochemistry applications rely upon, along with their corresponding double precision LAPACK routines are given in Table 1. These routines are discussed in greater detail in Refs. 7 and 3.

Eigen value and SVD computations are found in applications ranging from protein sub-state modeling to crystal growth analysis. Cholesky and LU solvers are found in applications such as determining molecular structure and molecular interaction in presence of a solvent.

While one may argue that these routines are only a "small'" part of the total application, we must not be deceived by them possibly being the applications' performance hotspots.

3. WHAT'S WRONG WITH LINEAR ALGEBRA ALGORITHMS ON CONTEMPORARY HARDWARE?

An algorithm that shows well on paper may not necessarily perform well on paper. The designers of the algorithms may not have taken into account all the factors that affect performance. For example, often the systems' memory hierarchy, Translation Look-aside Buffers, CPU architecture, compiler capabilities and functionalities are not taken into account by many mathematicians and computer scientists designing the algorithms, thereby leading to algorithms that falsely appear to be efficient, but fail to deliver the expected level of performance.

Table 1. Commonly used double precision LAPACK routines

Operation	Routines
Cholesky Factorization	DPOTRF
	DPPTRF
Cholesky Solver	DPOSV
	DPPSV
LU Factorization	DGETRF
LU Solver	DGESV
Least Squares Solver	DGELS
Singular Value Decomposition	DGESVD
QR Factorization	DGEQRF
Eigen Solver	DGEEV
	DSYEV
	DSYEVR
	DSYEVD

Another related problem is that complexity measures based on operation counts do not take into account data loading latencies, data locality, instruction and data pre-fetching and prediction. While in theory, it is possible to load data instantaneously, it is not achievable in practice on real hardware. Relying on such partial-view performance prediction methods leads to false assumption of the capabilities of an algorithm.

It is also often falsely assumed that an algorithm that perform well on one hardware platform and theoretically "proven" using complexity measures to be efficient, would also perform well on another hardware platform, as well as subsequent generations of hardware platforms. Algorithms and their implementations have to be reevaluated for their effectiveness, and redesigned if necessary.

Having a biochemistry software package relying on algorithms designed under these false views would lead to inefficiency of the software package. Building and using Grid computing environment to enable such inefficient software to deliver results quicker or enable finer analysis of a problem at hand, would be gross mismanagement of investment.

With linear algebra operations at the core of many biochemistry applications, we need to ensure that the numerical routines used are optimal.

4. PERFORMANCE TESTS

We conducted performance tests to benchmark the performance of the Cholesky solver, LU solver, QR Factorization, SVD, and Eigen solvers using routines from the OptimaNumerics Libraries on Intel Xeon EM64T, Intel Itanium II, and IBM PowerPC970 architectures. The benchmarks were conducted with no other load on the machines.

4.1. Intel Xeon EM64T

On the Intel Xeon EM64T (Nocona, IA-32E) architecture, the benchmarks were conducted on machines with Xeon EM64T CPUs running at 3GHz. There were 2 CPUs in the machine, but only 1 CPU was used. The compilers used were Intel Fortran Compiler version 8.1 and Intel C++ Compiler version 8.1. The matrices used were generated uniformly distributed random matrices. The memory available on the machine was 4GB of SDRAM. Each CPU has 12kB instruction L1 cache, 20kB data L1 cache, and 1024kB L2 on-chip cache.

4.2. Intel Itanium II

On the Intel Itanium II (IA-64) architecture, the benchmarks were conducted on machines with Itanium II CPUs running at 900MHz. There were 8 CPUs in the machine, but only 1 CPU was used. Intel Fortran Compiler version 8.1 and Intel C++ Compiler version 8.1 were used. The matrices used were generated uniformly distributed random matrices. The memory available on the machine was 15GB of SDRAM. The CPUs have 16kB instruction L1 cache, 16kB data L1 cache, 256kB L2 on-chip cache, and 1.5MB L3 on-chip cache.

4.3. IBM PowerPC970

On the IBM PowerPC970 architecture, the benchmarks were conducted on IBM eServer BladeCenter JS20 machines with PowerPC970 CPUs running at 1.6GHz. Only 1 CPU was used for each of the tests. Native IBM XL compilers were used on the machines running Linux. The matrices used were generated uniformly distributed random matrices.

4.4. OptimaNumerics Libraries: OptimaNumerics Linear Algebra Module

The OptimaNumerics Linear Algebra Module is part of OptimaNumerics Libraries. OptimaNumerics Linear Algebra Module provides a complete LAPACK implementation. The routines incorporated in OptimaNumerics Linear Algebra Module features algorithms which make efficient use of the CPU and memory available. In addition to exploiting the hardware features in the CPU, the algorithms take into account the memory architecture and processor architecture on the machine as well.

4.5. Benchmark Results

The results of the double precision benchmarks conducted are shown in Figures 1-17.

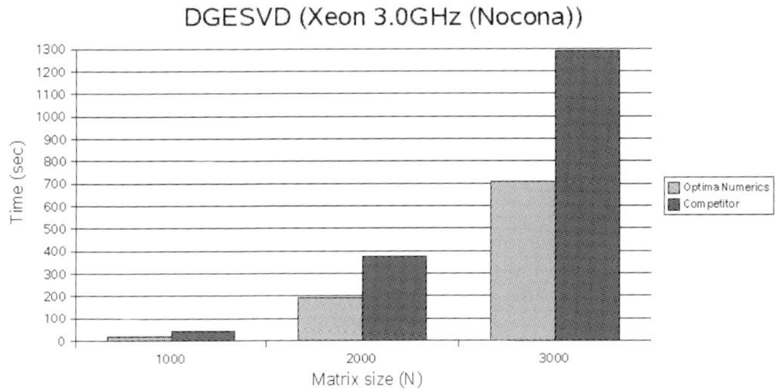

Figure 1. Performance of SVD routine from the OptimaNumerics Libraries compared to the closest competitor on Intel Xeon EM64T (Nocona) CPU.

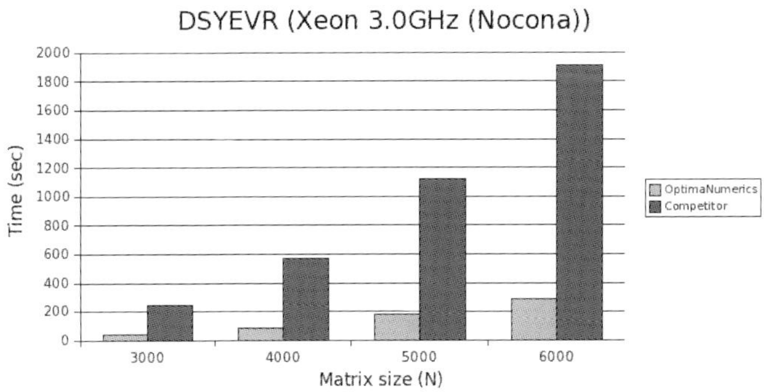

Figure 2. Performance of symmetric eigensolver from the OptimaNumerics Libraries compared to the closest competitor on Intel Xeon EM64T (Nocona) CPU.

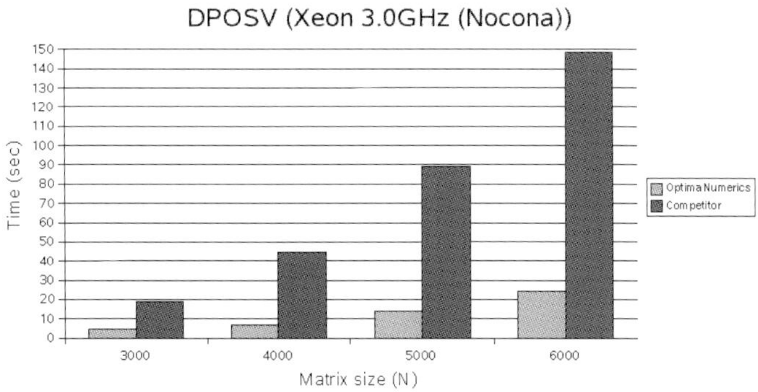

Figure 3. Performance of symmetric eigensolver from the OptimaNumerics Libraries compared to the closest competitor on Intel Xeon EM64T (Nocona) CPU.

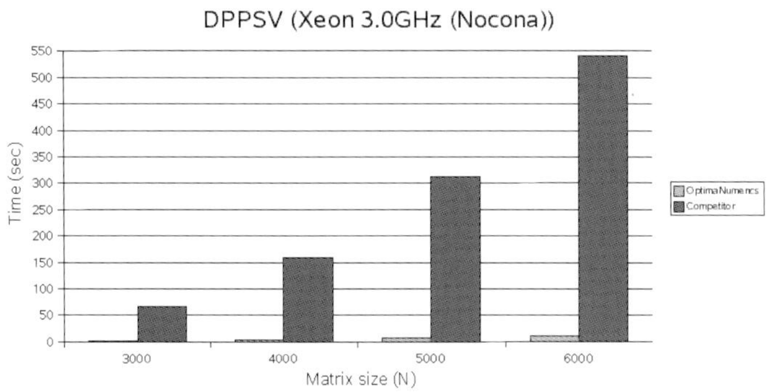

Figure 4. Performance of Cholesky solver from the OptimaNumerics Libraries compared to the closest competitor on Intel Xeon EM64T (Nocona) CPU.

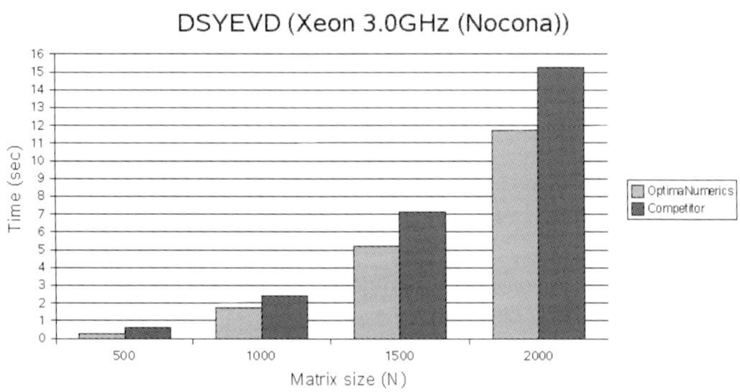

Figure 5. Performance of Cholesky solver from the OptimaNumerics Libraries compared to the closest competitor on Intel Xeon EM64T (Nocona) CPU.

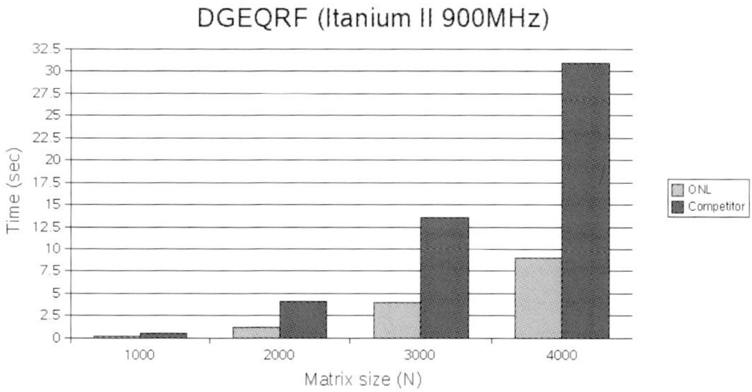

Figure 6. Performance of generalized eigensolver from the OptimaNumerics Libraries compared to the closest competitor on Intel Itanium II CPU.

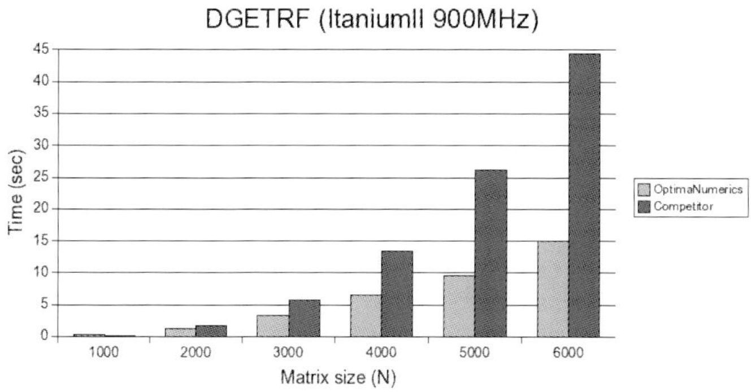

Figure 7. Performance of QR factorization routine from the OptimaNumerics Libraries compared to the closest competitor on Intel Itanium II CPU.

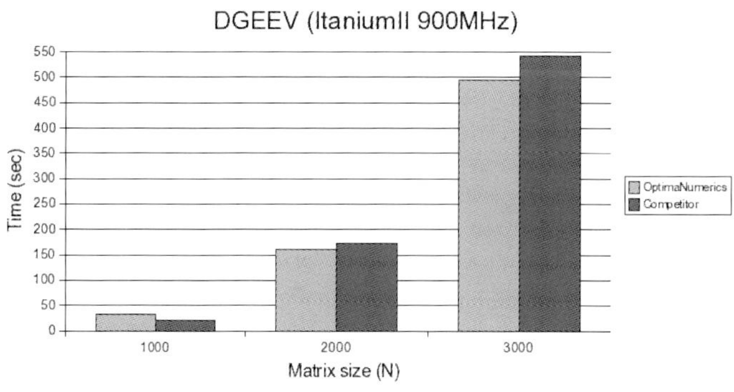

Figure 8. Performance of LU factorization routine from the OptimaNumerics Libraries compared to the closest competitor on Intel Itanium II CPU.

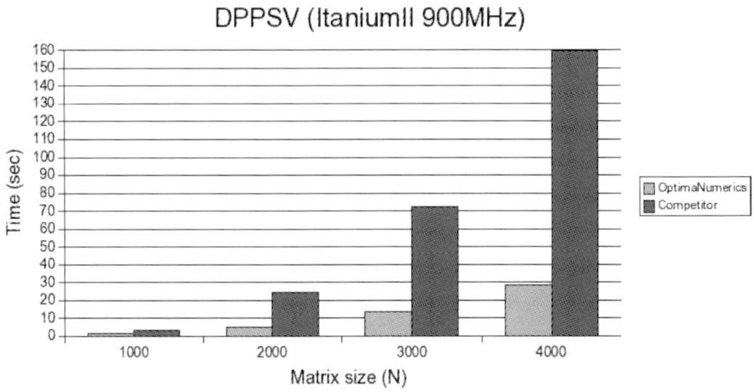

Figure 9. Performance of Cholesky solver from the OptimaNumerics Libraries compared to the closest competitor on Intel Itanium II CPU.

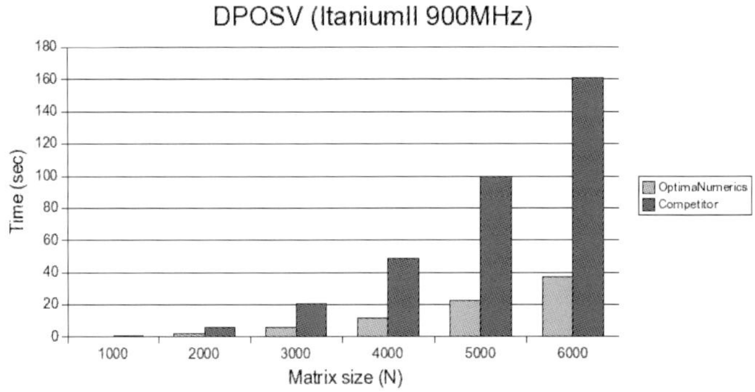

Figure 10. Performance of Cholesky solver from the OptimaNumerics Libraries compared to the closest competitor on Intel Itanium II CPU.

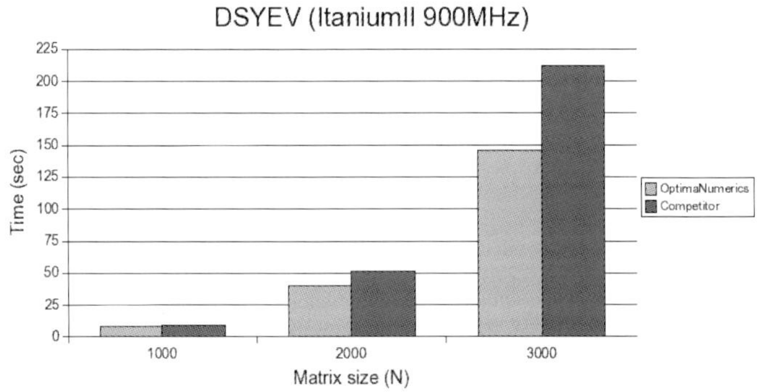

Figure 11. Performance of eigensolver from the OptimaNumerics Libraries compared to the closest competitor on Intel Itanium II CPU.

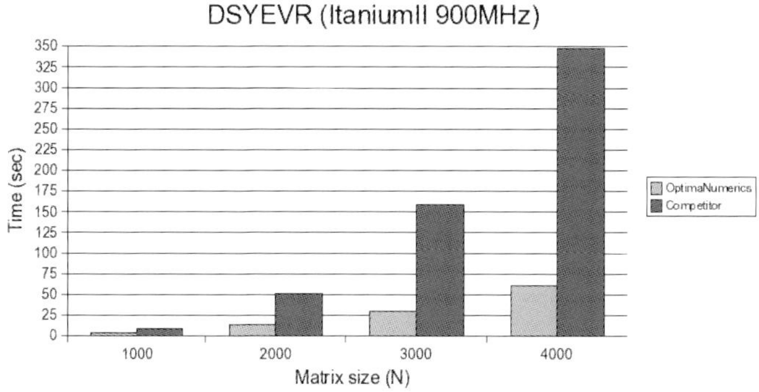

Figure 12. Performance of eigensolver from the OptimaNumerics Libraries compared to the closest competitor on Intel Itanium II CPU.

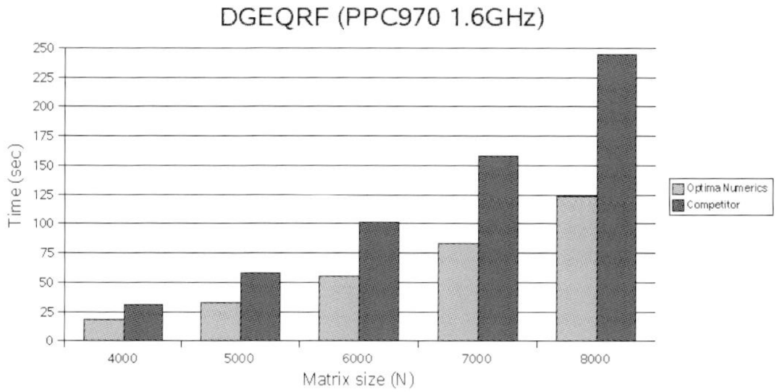

Figure 13. Performance of eigensolver from the OptimaNumerics Libraries compared to the closest competitor on Intel Itanium II CPU.

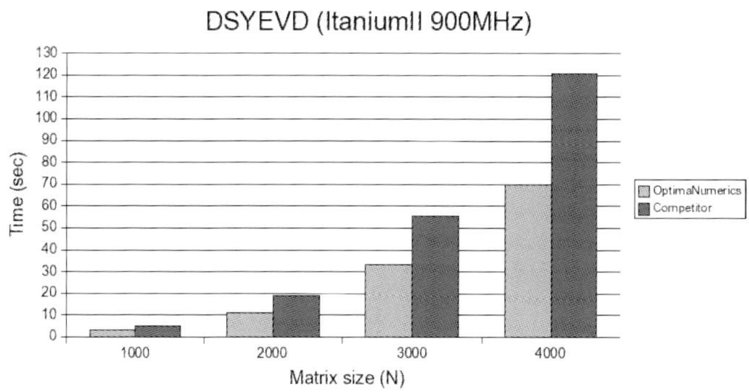

Figure 14. Performance of QR factorization from the OptimaNumerics Libraries compared to the closest competitor on IBM PowerPC970 CPU.

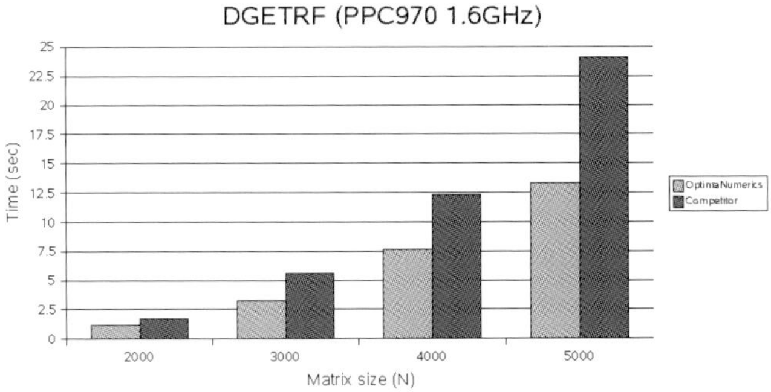

Figure 15. Performance of LU solver from the OptimaNumerics Libraries compared to the closest competitor on IBM PowerPC970 CPU.

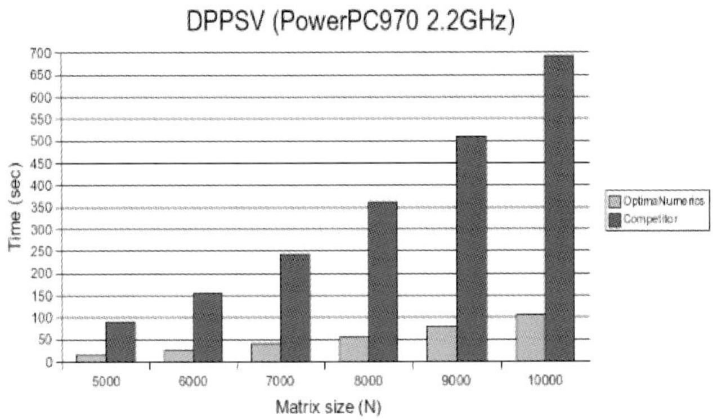

Figure 16. Performance of Cholesky solver from the OptimaNumerics Libraries compared to the closest competitor on IBM PowerPC970 CPU.

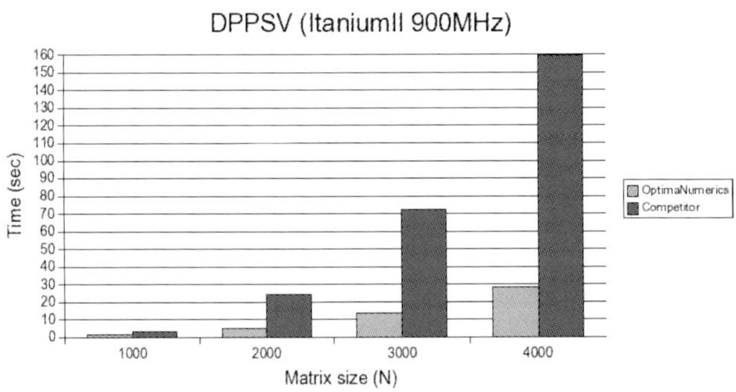

Figure 17. Performance of Cholesky solver from the OptimaNumerics Libraries compared to the closest competitor on Intel Itanium II CPU.

5. DISCUSSIONS AND CONCLUSION

As seen in the performance graphs, it is evident that LAPACK routines provided by the hardware manufacturers -- the closest competitors -- are under-performing, compared to OptimaNumerics Libraries.

It is to be noted that OptimaNumerics Libraries routines are implemented in high level languages -- C and Fortran -- rather than in assembly language. The code base is 100% portable.

We can therefore draw the following conclusions:

1. The extreme performance advantage delivered by OptimaNumerics Libraries is significant enough to affect the overall performance of biochemistry applications.
2. The large performance difference obtained using OptimaNumerics Libraries warrants a biochemistry application to take advantage of faster numerical tools, prior to committing large investments into Grid computing.
3. High performance can be achieved for a CPU-intensive computation problem, with code written in C and Fortran, using highly efficient, novel algorithms.
4. It is possible to achieve performance significantly higher than that attainable using hardware manufacturers' libraries.
5. While BLAS level code may be efficient, an implementation of LAPACK layered above the efficient BLAS is not guaranteed to be similarly efficient.

As shown in Ref. 10, efficiencies of scientific computing tools have great financial implications. In addition, we cannot assume that since Moore's Law states that performance doubles every 18 months, we can simply keep buying new hardware to achieve better performance. For example, since on the Intel Xeon EM64T CPU, the OptimaNumerics Libraries Cholesky solver is almost 50 times faster than the closest competitor, it will be more than 8.5 years (102 months) before the same level of performance can be achieved, assuming Moore's Law holds!

REFERENCES

1. Basic Linear Algebra Subroutines (BLAS). http://www.netlib.org/blas/.
2. Bientinesi, P., Gunnels, J. A., Gustavson, F. G., Henry, G. M., Myers, M. E., Quintana-Orti, E. S., and van de Geijn, R. A. The Science of Programming High-Performance Linear Algebra Libraries. In *Proceedings of Performance Optimization for High-Level Languages and Libraries (POHLL-02)* (2002), Association for Computing Machinery.

3. Golub, G. H., and van Loan, C. F. *Matrix Computations*, 3 ed. The Johns Hopkins University Press, 1996.
4. Goto, K., and van de Geijn, R. On Reducing TLB Misses in Matrix Multiplication. Tech. Rep. TR-2002-55, University of Texas at Austin, 2003. FLAME Working Note 9.
5. Gunnels, J. A., Henry, G. M., and van de Geijn, R. A. A Family of High-Performance Matrix Algorithms. In *Computational Science – 2001, Part I* (2001), V. N. Alexandrov, J. J. Dongarra, B. A. Juliano, R. S. Renner, and C. J. K. Tan, Eds., vol. 2073 of *Lecture Notes in Computer Science*, Springer-Verlag, pp. 51 – 60.
6. Linear Algbra Package (LAPACK). http://www.netlib.org/lapack/.
7. Moreland, T., and Tan, C. J. K. Performance of Linear Algebra Code: Intel Xeon EM64T and ItaniumII Case Examples. In *Computational Science and Its Applications: ICCSA 2005* (2005), O. Gervasi, M. L. Gavrilova, V. Kumar, A. Lagan`a, H. P. Lee, Y. Mun, D. Taniar, and C. J. K. Tan, Eds., vol. 3483 of *Lecture Notes in Computer Science*, Springer-Verlag, pp. 1120 – 1130.
8. Scalable Linear Algbra Package (ScaLAPACK). http://www.netlib.org/scalapack/.
9. Tan, C. J. K. Performance Evaluation of Matrix Solvers on Compaq Alpha and Intel Itanium Processors. In *Proceedings of the 2002 International Conference on Parallel and Distributed Processing Techniques and Applications (PDPTA 2002)* (2002), H. R. Arabnia, M. L. Gavrilova, C. J. K. Tan, and et al., Eds., CSREA.
10. Tan, C. J. K., Hagan, D., and Dixon, M. A Performance Comparison of Matrix Solvers on Compaq Alpha, Intel Itanium, and Intel Itanium II Processors. In *Computational Science and Its Applications: ICCSA 2003* (2003), V. Kumar, M. L. Gavrilova, C. J. K. Tan, and P. L'Ecuyer, Eds., vol. 2667 of *Lecture Notes in Computer Science*, Springer-Verlag, pp. 818 – 827.

SOLUTIONS FOR GRID COMPUTING IN LIFE SCIENCES

ULRICH MEIER
Sun Microsystems GmbH, Brandenburger Str.2
40880 Ratingen, Germany
Email: Ulrich.Meier@sun.com

Life Sciences researchers in industry and academia are faced with an exponential growth of data and the need to manage the data over their entire lifecycle to extract knowledge for new discoveries. This increases the collaborative challenge of dealing with redundant data sets and information lifecycle management, bandwidth issues, and shared resources. Grid computing allows for more efficient virtualization and provisioning of available resources. It enables virtual teams to collaborate on complex tasks beyond organizational boundaries, over secure networks. Recent installations like Bayer Healthcare in the pharmaceutical industry, the University of Antwerp and the Canadian Bioinformatics Resource in the academic space are introduced. These state-of-the-art grid deployments showcase the use of grid technologies to solve life science IT challenges.

1. LIFE SCIENCES CHALLENGES

1.1. Data Growth

During the last 15 years the content of biological data bases grew exponential. The growth of Genbank[1] is accompanied by an increasing number of databases and related biological sciences subfields describing very large-scale data collection and analysis. The number of new -omics fields is a good indicator for the progress and diversity in this area[2]. This all leads to a doubling of biological data every 12 to 15 month. With this, the current growth rate exceeds Moore's Law, which predicts a doubling in CPU speed every 18 month. Over the last decade, the increase in disk storage capacity has followed Moore's Law.

To keep up with the data volume and the subsequent data processing, pharmaceutical companies can either invest in IT infrastructure or use the existing resources in a more collaborative way by implementing Grid technologies.

---

[1] http://www.ncbi.nlm.nih.gov/Genbank/genbankstats.html
[2] http://www.genomicglossaries.com/content/omes.asp

1.2. Stability of Data Definitions

A second trend, which seems to lead in the opposite direction can be derived from the statistics of the Human Protein Index at the EBI[3]. Although continued research is being undertaken, the number of entries decreases over time. Redundant or wrong entries have been eliminated but also a shift in definitions has occurred. Entries that have been named individual proteins are now regarded as a single complex. These fluctuations lead to a greater need of storing experimental data and repeating a numerical analysis with the latest set of databases.

2. GRID COMPUTING

2.1. Definitions

Grid technology allows disparate systems to be pooled and managed as a common computing resource. It is a way to optimize the throughput of workloads across the computing infrastructure and provide access to resources that were previously unavailable[4]

Table 1. Grid definition matrix[5]

	Sun	IDC	Gartner
Compute	Desktop Grid	Hallway Grids	X
	Cluster Grid	Campus Grids	X
	Enterprise Grid	Intraprise Grids	Computing Grid
	Global Grid	Power Grids	
		Provisioning Grids	
		Internet Computing Clusters	
Data	Data Grid	X	Data Grid
Visualization	Visualization Grid	X	Collaboration Grid
Access	Grid Access	X	X

1. Grid is the natural evolution of distributed computing
2. Grid is a business model and an operational concept
3. Grid is a set of technologies that affect the architectural design of IT infrastructure and deployment
4. Grid is a combination of hardware and software that distributes jobs across the IT infrastructure

---

[3] http://www.ebi.ac.uk/IPI/IPIhuman.html

[4] Foster, I. & Kesselman, C. (Eds). The Grid: Blueprint for a New Computing Infrastructure. Morgan-Kaufmann (1999)

[5] http://www.theopensourcery.com/osrevGrids.htm

2.2. Collaborative Research Challenges

A typical Collaborative Research Environment in a big pharmaceutical company may look like this: A research center that has its own data, applications and algorithms. Interaction with headquarters, system integrators, contacted research organizations (CROs) and informatics partner are necessary. A collection of public, in-house and private 3rd party data bases has to be accessible by the researchers. Furthermore the databases have to be updated on a regular bases, in-house copies have to be maintained (an Internet searching using a novel sequence can result in the loss of the patent!!), the databases have to be kept coherent, redundancies have to be eliminated and inconsistencies and wrong entries (this is the most difficult one) have to be cleared.

This is even more complex for a multitude of workgroups and different departments with different research aspects and a different nomenclature resulting in different ontologies (definitions).

Sun's Java Enterprise System software enables collaboration between groups and organizations. The times of mergers and acquisitions are not over (Sanofi – Aventis merger in Summer 2004)

Directories, Identity Services, Java Card, Secure and Mobile Portals, Calendar Services, JDS, Trusted Solaris, and N1 Service Provisioning System build the collaborative software platform to manage these challenges. These middleware services have to complement each other, have open interface and the flexibility to be exchanged with service from third parties.

2.3. Deployment Architecture Bio-ClusterGrid

The Bio-ClusterGrid[6] is the first of many deployment architectures that realizes the benefit of the BioBox initiative. Implemented using Sun's Web Start Flash technology, the core components of the Bio-ClusterGrid are:

- Solaris operating environment
- Ready-to-use bioinformatics applications
- Sample databases
- Sun Grid Engine (SGE)
- Grid Engine Portal (GEP)

---

[6]https://biobox.dev.java.net/

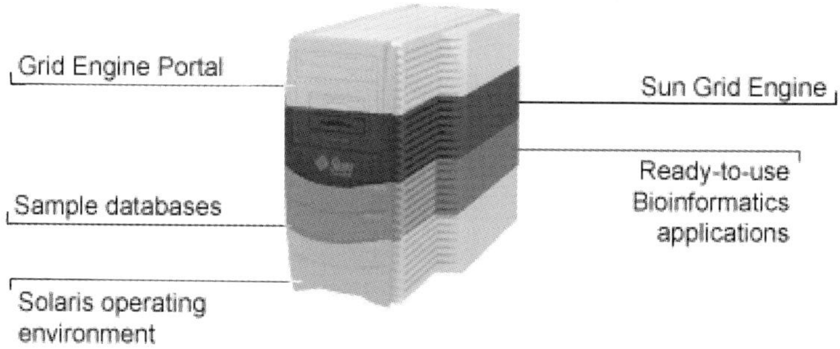

More than 20 of the most popular bioinformatics applications are made available on the Bio-ClusterGrid through the portal, which greatly enhances application usability. Biologists access the portal either through a browser in Sunray Thin Clients in the access tier or through any browser enabled device. The Grid Engine software provides the resource management mechanism to schedule all the bioinformatics applications to run on the cluster of execution servers.

3. UTILITY COMPUTING

The goal of the Utility Grid is to create ubiquitous, standardized computing, with simple, transparent pricing. This is a natural outgrowth of grid technology – based on open, tried and true products/architecture that enables "The Network is the Computer." motto. As the Utility Grid products evolve, they may support additional operating systems (in addition to Solaris) in the future (i.e., Windows and Linux). True utility computing is technologically complex, but simple to use – just plug it in and mitigates complexity in the data center. With Solaris 10, Nauticus, and Identity Management solutions, we have the ability to secure the network, using federated identities that move throughout the compute and storage system allowing the composition of a single isolated virtual system for a user.

One last, and very important point is that we have started by building to the most demanding marketplace -- the Internet ... knowing that techniques that scale, secure, and remain available at this order of complexity, will naturally be easy back-fits into the data center.

Top industries and tasks for Sun Grid

- When we look at the various industries that are responding to challenges for computing resources in your organizations, you see a lot of momentum towards distributed computing.

- In the financial services industry, risk and portfolio analysis, pricing, and Monte Carlo simulations are common applications that fit this model.
- In the oil and gas industry, we see companies collecting data worldwide in geographic locations ranging from headquarters to field offices to drilling platforms at sea. They have distributed data and they are trying to collaborate and share with those using centralized data resources. It is an interesting challenge that lends itself to distributed solutions.
- The electronic design automation industry, particularly chip design, is an area that is been an early adopter of distributed computing solutions for simulation, verification, and regression testing. In digital content creation (e.g., creation of digital effects including commercials and films), frame rendering is another area where distributed computing technologies have been adopted.
- In the automotive field, crash testing, stress testing, computational fluid dynamics (e.g., determining the airflow over vehicles) lend themselves very well to distributed computing solutions.
- In the government and educational/research arenas, weather analysis and image processing are among several applications that are applicable here.
- In the health/life sciences industry ... we've mentioned genomics already. But certainly proteomics and most of the "-omics" fields are other areas that are faced with exponentially large and growing requirements for computing power and large databases – along with medical imaging and new drug simulations.

4. USE CASES

4.1. Bayer Healthcare

The pharmaceuticals division of Bayer HealthCare AG develops and delivers high-quality medications to treat emergent diseases such as cancer, obesity, and diabetes, as well as cardiac, urological, and circulatory disorders. Recent products are Adalat, Avalox, Ciprobay, and Levitra. Developing these medications requires enormous computing resources, particularly for molecular modeling. To provide this level of computational performance, the pharma division of Bayer HealthCare chose to expand its infrastructure with a Linux cluster from Sun Microsystems[7].

Andreas Göller, Ph.D., senior research scientist for chemical research in the pharma division, says the computational chemists and scientists had previously worked on SGI workstations and servers because of their need for graphics

[7] http://www.sun.com/solutions/documents/articles/ls_bayer_aa.xml

capabilities. However, the division's rapidly ballooning need for computing power made it apparent that the company had to do something to improve its cost-to-performance ratio. Sun's strength in the life sciences made an impression as well. "We chose Sun over other Linux providers because the company has a very strong presence in all research sectors, especially in the area of bioinformatics," Göller said in a statement. "Our confidence in the investment security offered by Sun products also played an important role in our decision."

The Bayer HealthCare Linux cluster runs N1 Grid Engine (formerly Sun Grid Engine) software. Two Sun Fire V65x servers, acting as management nodes, are the basis of the cluster and 100 Sun Fire V60x servers, each with two processors, serve as the cluster's computation nodes. The system also includes two Sun StorEdge SE 3510 memory units with a Sun TapeLibrary and an additional Sun Fire V65x server as backup.

Science + Computing[8], a Sun solution partner based in Tübingen, Germany, implemented the cluster. Building on its focus on the technical computing market, Science + Computing has been a leading provider of professional Linux clusters in Germany and integrated the first industrial Linux cluster for DaimlerChysler in 1999.

As part of implementing the cluster environment tailored to Bayer's need, Science + Computing set up the hardware, installed the cluster, and integrated the workstations. The system integrator also optimized computational chemistry applications from Accelrys, Schrödinger, and Tripos, as well as Bayer's own in-house applications, on the Linux platform. Applications vary from quantum mechanical calculations to library design, pharmacophore analysis, and *in silico* screening. According to Göller, Science + Computing's expertise saved Bayer the time and expense of hiring technical staff and establishing the new Linux cluster environment in-house.

4.2. CalcUA

On 14 March 2005 CalcUA, the most powerful computer cluster in Belgium was inaugurated officially at the University of Antwerp[9]. This opens a host of new possibilities for scientific research including the areas of mathematics, informatics, linguistics, biomedical sciences, physics and chemistry. CalcUA is built using computer systems provided by Sun Microsystems Belgium, located in Zaventem. In order to achieve speeds of two trillion calculations per second, i.e., 2 teraflops, 256

[8] Science + Computing AG: http://www.science-computing.de
[9] http://be.sun.com/press/releases/index.html

nodes will be installed. The nodes each have two 64 bit AMD processors and have a fast Myrinet connection. The installation is completed with 18TB of raid-storage.

CalcUA is an interdisciplinary project between several departments at the University of Antwerp in which mathematicians, informaticians, linguistic specialists, biomedical scientists, physicians and chemists are participating. Complex calculations, which at present might soon take one year, now can be reduced to a few hours using the computer cluster. The CalcUA is transforming the University of Antwerp into a Center of Excellence. As opposed to Bayer's industrial research environment, where only a few applications mostly from commercial vendors are present, CalcUA hosts a multitude of development environments and university developed highly specialized scientific software.

This opens a wide perspective of new and ground-breaking research horizons. Through computer simulations, for instance, Purkinje cells, the nerve cells of the cerebellum, can disclose their secrets. These *in silico* experiments will contribute to the research that tries to pinpoint the function of the cerebellum.

Physicians are investigating very small nanostructures such as quantum particles that are applied in lasers and carbon nano tubes which are used as components of electronic transmitters at nanometer level. The supercomputer will be used in chemistry to find out more about plasma for gas discharges applied in films to cover materials and to produce flat plasma screens.

In addition, the supercomputer will provide services in searching methods to reconstruct complex 3D objects through indirect measurements such as gas supplies in the surface or a cancer tumour in the human body, as well as in optimizing complex networks like a traffic or data network. The informatics department will have the opportunity to better support other research teams, for instance in defining the geometric structure of biomolecules such as proteines, or in detecting and comparing toxicological data via micro-arrays.

Linguistics research will evolve through the application of self-learning techniques using artificial intelligence. As such computers will be able to generate knowledge from text.

Ten percent of the supercomputer capacity will be reserved to the industry. Commercial collaboration with the industry has not yet been made concrete but contacts are being established with the petrochemical and pharmaceutical sectors. Via inter-university science groups, other universities in Belgium or abroad, can be partners in discovering the benefits of CalcUA.

4.3. Other Use Cases

4.3.1. *Plexxikon*

Sun provided a complete, reliable and massively scalable solution that can help Plexxikon[10] develop novel drugs from its Scaffold Discovery Factory more quickly and cost-effectively than was possible in the past.

The Scaffold Discovery Factory is a highly scalable, available computational platform infrastructure for drug discovery. It combines state-of-the-art technologies from a variety of disciplines, including co-crystallography, low-affinity biochemical screening and synthetic chemistry to rapidly discover new chemical scaffolds and high-quality drug leads.

4.3.2. *Oxford Glycosciences*

OGS[11] (now part of Celltech) has developed what it believes is one of the most powerful and sophisticated proteomics/genomics data factories in the world. It is managed almost seamlessly through a Sun grid computing infrastructure.

The key business challenges are:

- Decrease turnaround times during peak use
- Achieve better resource utilization
- Lower computing costs

4.3.3. *University of Calgary*

The University of Calgary implemented Sun technology[12] as the basis for its unique and powerful bioinformatics research environment. Scientists gain new insights by immersing themselves in a 10 feet by 10 feet CAVE (built by Fakespace Systems of Kitchener, Ontario) where they are surrounded by 3D stereo images of structures such as molecules, cells, or tissues. Hardware and software from Sun, along with support rendered under Sun's Center of Excellence for Visual Genomics program, position the university at least 2 years ahead of competing research organizations.

The key business challenges are:

- Gain competitive edge in bioinformatics research
- Provide immersive environment capable of conveying new insights into complex 3D structures

---

[10] http://www.sun.com/solutions/documents/success-stories/ls_pexxikon_bb.xml?&facet=-1
[11] http://www.sun.com/solutions/documents/success-stories/ls_oxford_bb.xml?&facet=-1
[12] http://www.sun.com/products-n-solutions/edu/commofinterest/compbio/

- Manage vast quantities of data quickly and easily
- Decouple the application development environment from the execution environment

The key business solutions are:

- Two-year head start on competitive undertakings
- Immersive 3D stereo environment provides unique visualization powers
- Capacity to store and process many terabytes of bioinformatics data
- Java 3D technology permits CAVE applications to be developed without consuming CAVE resources

4.3.4. *The Canadian Bioinformatics Resource* (*CBR*)

The Canadian Bioinformatics Resource (CBR)[13] provides biologists around the world with access to bioinformatics applications, large-volume data storage, basic training and help desk support. Currently, CBR[14] provides this service to National Research Council (NRC) scientists and academic/not-for-profit users associated with a university, hospital or government department. CBR resources are available for education and not-for-profit bioinformatics research purposes. CBR is currently running a pilot Industry program in conjunction with the British Columbia Institute of Technology (BCIT) and Vitesse Re-Skilling Canada to develop strategies to support business users in the near future.

5. CONCLUSIONS

Long used by scientific and engineering organizations in the Life Sciences industry, Grid Computing has now moved into the business mainstream as a proven technology that helps IT infrastructure contribute directly to business goals.

The evolution of grid has been ongoing for many years and so it is not new technology. Today, over 50% of Sun Grids implemented are commercial enterprises, going beyond the academia and research communities. While deploying a grid is not conceptually difficult, it is not rocket science either. The software is one small part of designing and implementing a total Grid Computing Solution. Utility Computing is the next evolutionary step and we should expect to see more of pay-per-use services in near future.

---

[13] http://ca.sun.com/en/press/archives/20021029.html
[14] http://cbr-rbc.nrc-cnrc.gc.ca/index_e.php

While classic Grid Computing has the flexibility and dynamic allocation of resource and workload, the provisioning of cost is still static. Therefore it is independent of the actual load on the grid. Utility Computing shifts this paradigm to a flexible and dynamic cost allocation to the next evolutionary level.

STREAMLINING DRUG DISCOVERY RESEARCH BY LEVERAGING GRID WORKFLOW MANAGER*

ANIRBAN GHOSH

Life Sciences, Infosys Technologies Ltd., Plot No. 44 & 97A, Electronics City, Hosur Road Bangalore, Karnataka 560100, India
Email: ghosh_anirban@infosys.com

ANIRBAN CHAKRABARTI, DHEEPAK R.A., SHAKEB ALI

Software Engineering and Technology Laboratory, Infosys Technologies Ltd.
Email: {Anirban_Chakrabarti, dheepak_ra, shakeb_ali}@infosys.com

Presently, pharmaceutical companies are under pressure to speed up time-to-market and reduce costs of producing novel drug molecules. One of the key enablers, which can be leveraged in various silos of this industry, to enhance productivity is grid computing. Grid allows linking up as many processors, storage and/or memory of distributed computers to make more efficient use of all available computing resources to solve large problems quickly. The benefits of grid computing include cost savings, decrease time to deliver results, enhance collaboration and harness existing computing resources. We have developed a workflow solution on grid called Discovery Research Workflow on Grid (DRWG) which helps automate and accelerate gene discovery research. DRWG enables the user to custom design a pipeline of compute intensive tasks which can be executed on heterogeneous platforms. The repetitive and complex set of tasks are efficiently managed by Grid Workflow Manager (GridWorM) which schedules, load levels and delegates tasks onto the available computing resources. This paper will discuss how high throughput functional annotation of genomic DNA sequences or profiling groups of protein sequences can leverage grid computing services.

1. INTRODUCTION

1.1. How Genomics Help Drug Discovery?

Presently, pharmaceutical companies are adopting targeted and rational drug discovery route. Traditional novel drug discovery programs were essentially based

*This work is partly supported by Life Sciences - Drug Discovery Informatics and Software Engineering and Technology Labs solution grants.

on the screening of a library of chemical compounds developed by a combinatorial chemistry approach. For this, a target protein molecule was essential for designing a drug. Only a small number of specific drug targets were available earlier. With the availability of human genome information, more targets have been identified, largely predicted by bioinformatics tools and later validated by molecular biology experiments. Moreover, the targets identified using the bioinformatics approach is more 'reliable'. This not only lead to a fewer number of drug candidates for pre-clinical testing, but also reduce the attrition rate of molecules in the clinical validation phase. The translation of genomics knowledge into drugs has conclusively established the importance of use of informatics in pharmaceutical research.

From the complete genome sequences of 261 organisms[1], there is immediate imperative to extract genuine new insights and discoveries from the genes and proteins that are coded for an organism. Bioinformatics has the potential to reduce cost and complexity of drug discovery projects and identify specific, selective targets and drug-able leads for a therapeutic program.

1.2. Grid Services Opportunities in Drug Discovery

There are many examples in drug discovery which are currently leveraging data and compute grid services. Some of the salient examples are: a) high throughput screening for lead molecules against targets for cancer, b) fightaids@home by executing auto-dock to screen for suitable inhibitor to active site of HIV protease[2], c) identifying protein profiles from tandem mass spectrometry from serum samples of Thalassemia patients, d) simulating complex cellular models, e) determining statistical trends from profiles of micro-array gene expression, f) computational protein folding modeling, and g) predicting protein-protein interaction in a biochemical pathway.

Discovery research in biological and chemical sciences involves repeated execution of compute intensive applications like BLAST[3], ClustalW[4], or HMMER[5] which form a part of the pipeline of tasks. In other cases, these research problems are solved by conducting a long computer simulation of a molecular system.

Each one of the research processes is executed on dedicated machines which results in long idle time and hence inflated budget. Either the manual intensive tasks are managed by trained specialists who are expensive, or scientists supervise the execution of the workflows resulting in ineffective use of the scientist time. In addition, there are overheads in terms of preparation time to initiate application and assimilation and presentation of scientific outcomes, which are all susceptible to human error. These challenges in discovery research can be suitably overcome by deploying grid services and help determine novel drug molecule fast and accurate.

1.3. Computational Workflows for Genomic Research

Quite unlike generic transactional workflows, computational workflow integrates a pipeline of scientific data management tasks. To accelerate and automate a complete research process, compute and data intensive workflow management provides speed, throughput, compute resource utilization, multi-application integration, diverse database access, and enable scientific collaboration. Shown in Figure 1, high level research processes for target identification and lead optimization. The computational workflows to each can be formed by integrating the computational tasks underlying the process maps.

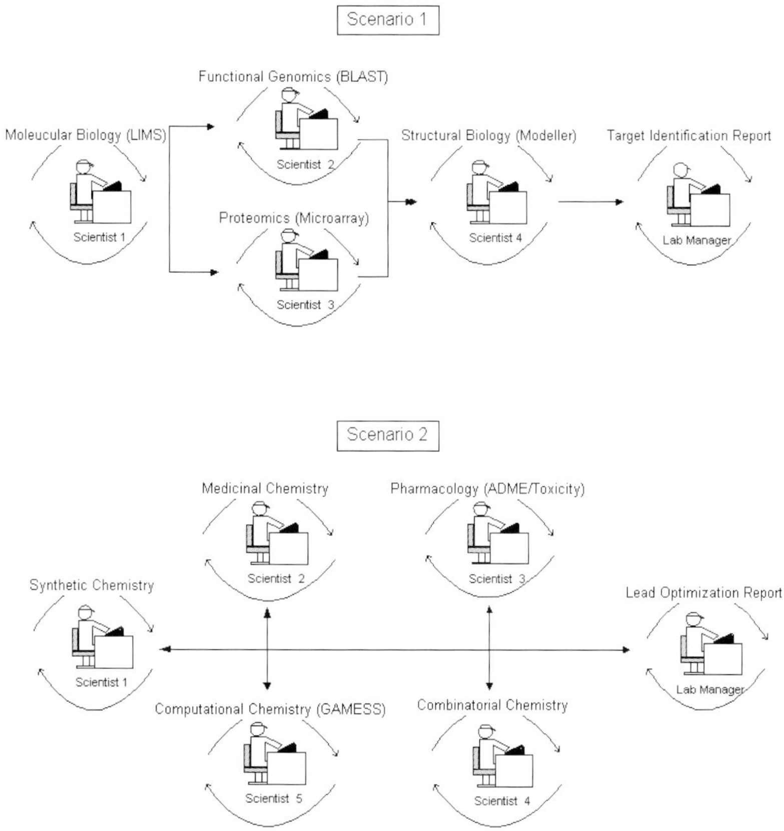

Figure 1. Representative target identification and lead optimization workflows.

1.4. Grid Services for Computational Workflows

Grid Computing[6] is defined as a mechanism to overcome heterogeneity of computing elements, operating systems, and also in terms of policy decisions and environment. A long-term vision of enterprise grid computing community is non dedicated seamless interoperability of different disparate systems which may be part of the same organization or different organizations. Grid computing is looked upon by many experts as a technology that can potentially change the world, like the Internet did. However, from the user's point of view, grid is nothing but a computer with huge amount of computing resource. There are workflows in the functional genomics[7] domain and other silos of drug discovery research, which require grid services, to run compute intensive jobs through a workflow. Therefore, there is a need for process level workflow definitions to interact with the underlying heterogeneous grid infrastructure wherein the jobs comprising the workflow can be distributed across the infrastructure. This virtualization and load balancing results in improved efficiency, as the same infrastructure would support more load and hence lowering the overall total cost of ownership.

2. METHODS

2.1. Genomic Sequence Analysis for Target Identification

Sequencing projects obtain short nucleotide sequences or *Expressed Sequence Tags (ESTs)*[8] which are mapped for its chromosomal location and putative gene function. ESTs have applications in the discovery of new human genes, mapping of the human genome, and identification of coding regions in genomic sequences. Determination of biological functions of each string of DNA sequence is important to understand its biological context e.g., whether they are involved in horizontal gene transfer, whether a cluster of protein sequence belong to microbial genomes, whether a collection of gene sequences belong to same biochemical pathway of the genome. All molecular findings are an important first step to identifying a drug target in any drug discovery research program.

The NCBI GenBank[9], RefSeq BLAST[10], SWISSPROT[11] and other data bases are updated with functional DNA, RNA and protein sequences. Most of these databanks grow at exponential rate. As of April 2004, there are over 44,575,745,176 base pair of DNA found in 40, 604, 319 sequences in the GenBank. Upon querying for similar nucleotide sequence from these databanks, the commonly occurring function description for all the matched sequences is studied. Once the function to the as yet unknown nucleotide sequence is assigned – the DNA sequence is annotated with a biological function. The same is true with protein sequences, one finds the similar protein sequences and all the homologous sequences are studied for common traits or a sequence profile is generated. Thus a whole suite of experiment and computational analysis is performed to annotate a gene or profile a protein sequences in the field of functional genomics.

2.2. Computational Workflow for Sequence Analysis

The solution consists of a workflow builder, wherein the pipeline of applications is loaded together with the input data. Each of the individual tasks in each workflow manage compute intensive algorithms, large scale data processing scripts, statistical parsing routines, web based query, or invoke a third party application. Once the task is compiled, the Grid engine schedules and manages various tasks. Upon completion of all tasks, a final report is submitted for the end user's analysis.

The analytical workflow (refer to Figures 2a and 2b for the screenshot of the workflow GUI and the screenshot for the BLAST input) starts by submitting nucleotide sequence and issues it as a query to the NCBI (National Center for Biotechnology Information) sequence databases. The sequence alignment application BLAST (Basic Local Alignment Search Tool) executes a massive search to retrieve a set of homologous sequences. These hits are screened for high percentage similarity and identity as can also be identified through the score and e-value of the output. Select accession numbers are used to issue a query to retrieve the PubMed Ids identifying articles relating to the homologous sequences. The same query is issued for which source organism does this gene belong and functional role this sequence performs. A report is created for the list of highly probable gene sequences that the unknown sequence is mostly likely to be similar. A summary note is evaluated to highlight the statistics of the outputs, e.g. how many hits belong to human source? The job sequence ends by clustering a few screened gene sequences with the unknown sequence using a multiple sequence alignment application ClustalW. This application best aligns all the sequences to understand the detailed feature of the sequence domains. A few simple statistical routines are used to perform roll up statistics on the aligned sequences and the parameters from the output are correlated to obtain satisfactory confidence on the data output. A final report is generated as to which gene family the unknown sequence most likely belong, with sufficient statistical notes to highlight the biological significance of the nucleotide sequence.

The novel features of the workflow include scheduling and distributing tasks from various sections of the workflow to the available computing platforms. The tasks are distributed and monitor for their quality and timeliness of services. It is also designed to deliver a custom designed report for the entire job, after running suitable parser and screening scripts running each of the modular tasks. In summary, this computing workflow will enable integration of multiple applications and improve the flow of data and deploy the tasks effectively on the distributed computing infrastructure.

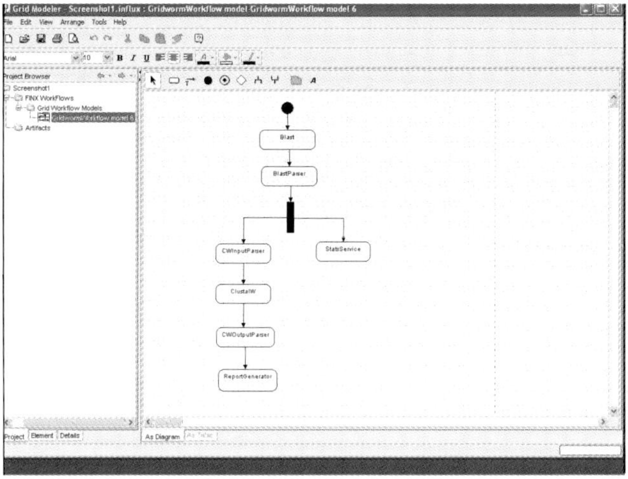

(a)

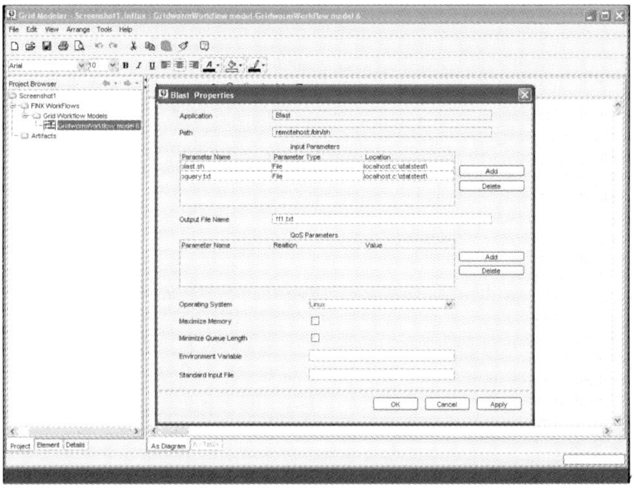

(b)

Figure 2. (a) Screenshot of the GridWorM interface. (b) Screenshot for entering parameters to BLAST.

2.3. Grid Workflow Manager (GridWorM)

The **Grid Workflow Manager (GridWorM)** allows the user to submit the jobs through a workflow. The workflow allows the integration of applications with enterprise entities like web services, relational data bases, and decision making.

2.3.1. *Technical Features of GridWorM*

GridWorM provides the user with the interface to provide relationships among the jobs. The features that GridWorM support are:

(a) Jobs Submissions: Users can submit jobs using a Graphical User Interface developed as part of the GridWorM application. The user can submit jobs based on the BPEL specifications, which has become an industry standard now.

(b) Job Relationships: Users can provide complicated conditional relationships among the different jobs.

(c) QoS Support: The GridWorM workflow manager provides variety of QoS support including availability, and trust levels[12].

(d) Infrastructure Support: GridWorM supports infrastructure level support in terms of making queries to multitudes of databases like MySQL, Oracle (different versions), and Postgres. Later versions of GridWorM will also support integration with enterprise level messaging services like JMS and MQ.

(e) Web Services integration: Jobs can either be standalone applications or web services. Standalone applications can also interact with Web Services seamlessly.

2.3.2. *GridWorM Architecture*

The inter-relationship between different components of GridWorM is shown in Figure 3. A brief description of each is described below:

GWLang: The GridWorM language is specifically designed for the workflow in mind. It is based on XML and has properties of Business Process Execution Language (BPEL) and Grid Services Flow Language (GSFL). GWLang combines the advantages of both BPEL and GSFL in a scalable manner. It inherits the relationship and QoS models from BPEL and it inherits the grid services model from GSFL. In addition, it also supports standalone applications, file management, and infrastructure level support like native database queries, opening remote shell (rsh) or remote copy (rcp) facilities. The unified grid model necessitated the development of the GWLang language which provides features unavailable in any of the existing job flow or business flow languages.

GWLang Generator: Another important component of the GridWorM is the GWLang parser. It is responsible for converting the user requirements from the GUI to the XML based GWLang language. The language is not exposed to the external users; however the user may choose to enter their requirements through the native GWLang also. The parser also converts the user requirements specified through other workflow specification language like BPEL. The parser is developed using Apache XMLBeans[13] which allows the developer to access the full power of XML in a Java friendly way.

GridWorm Pre-parser: Pre-parser is added because of mapping dynamic Web Services and JDBC calls in GWLang.

GridWorM Manager: GridWorM manager manages the different state machines within the GridWorM. GridWorM manager receives the jobs from the GWLang pre-parser and instantiates a GridWorM state machine. It also generates a unique workflow before giving it to the state machines.

GridWorM State Machine: The state machine in GridWorM manages the states and relationships among the different jobs within a particular workflow. It uses the Web Service provided by MAGI to submit the job. It continually polls for the status of the submitted job, based on the submission id returned by MAGI.

Guided User Interface: The user can use the tools provided on the interface to create the workflow and upload relevant input data. Each of the applications can be loaded with its associated parameters on the workflow. Once the computational workflows are saved, the can be resubmitted with minimum or no changes.

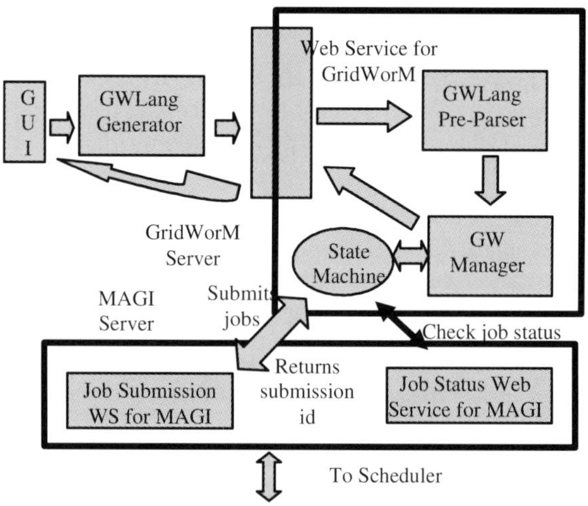

Figure 3. Schematic for GridWorM Architecture.

3. RESULTS

3.1. Significance of the Case Study Reports

For each workflow submitted to the GridWorM a complete report is obtained as highlighted in Table 1. The abbreviated report on a fungal COX-2 protein sequence depicts that after searching the nr-db, it found 201 sequences to have exactly 51-60% sequence identity. The parser output identified 9 hits from another fungal family and selected all protein sequences better than 70% sequence for multiple sequence alignment. Upon pair-wise alignment of all sequences, a common profile is depicting in the report along with distances in the phylogenetic map. To summarize, this reporting facility can be tweaked to produce gene or protein family classification, in the way suitable for the scientists to derive significance from the raw data sets.

Table 1. Report highlighting the various data obtained executing various tasks of the workflow.

Sections of the Report	Representative Data
Report ID: Date: User: Title	192.168.206.99: Mon Apr 11 2005 10:00; Dr. Anirban Ghosh; Report for annotating the Nucleic Acid/Protein Sequence
Details of the query nucleotide/protein sequence	sp|P00411| COX2_NEUCR; Cytochrome c oxidase subunit2; EC 1.9.3.1; Neurosporacrassa.
Program Detail: Package Used: Database	Blastall: BLASTp: nr-db
Closely lying sequence - Nucleotide ID: Protein ID: Length: Score: E-value: Identity: Start Seq: End Seq: Organism: Sequence	NP_074950.1:P20682: 250:476:e-133: 91%: 1: 250: Podospora anserine : MGLLFNNLIMNF
Simple Statistical Analysis: Total no. of HITS: Total no. of HITS between 71-80%: Quarter Percentile	250: 6: 0.0
Top three organism which has maximum number of hits	Candida Glabrata – 9
Sequence with identity greater than 70% included to generate a profile by ClustalW	gi|117030|sp|P00411|COX2_NEUCR; gi|12408617|ref|NP_074950.1;
Multiple Sequence alignment	MFFLINKLVMNLLNQVSVFINR ***:**.:*** * *:::::: .: .:***:***:*

3.2. Performance

The system is tested using GridWorM 1.0 and MAGI 1.0[12]. The prototype version of GridWorM has been developed to interact with the MAGI 1.0 system to provide the desired end-to-end result to the user. The GridWorM system interacts with the MAGI system through Apache Tomcat 4.1. Final scheduling is handled by the CONDOR[14] system.

We define the performance metrics for our study as Average Peak Utilization. We take the average of the utilizations of different machines for the period when all the machines were busy i.e., the grid system was running at its peak capacity. We needed to devise this metric as the traditional average CPU utilization will include the times when some machines were idle due to lack of jobs in the system.

The metric was tested on 2 dual proc Xeon™ (2.8 GHz, 1 GB) and a single proc (3.2 GHz, 1 GB) Dell server connected within a 100Mbps LAN. The results were based on the 25 workflows, each consisting of applications like BLAST, ClustalW and their parsers and report applications. The average peak utilization (in percentages) is 99.5% on machine 1, 96% on machine 2 and 88% on machine 3, as shown in Figure 4. This figure shows that the underlying grid infrastructure is utilized maximally and uniformly. The unequal utilization of different machines is because of the wide difference in execution times of the applications run through the GridWorM workflow manager. On a single processor 3.2 Ghz machine, the time taken to run 25 workflows is around 8.5 hours. It took a total time of 2.55 hours to execute 25 workflows submitted to the three machines. Therefore, the GridWorM along with MAGI shows a gain of 67% in terms of elapsed time.

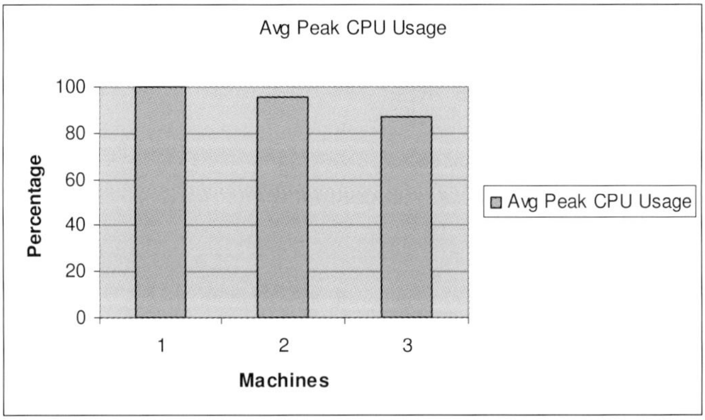

Figure 4. Plot of Average Peak usage of CPU of test machines.

Another metric for performance of a GRID based workflow solution is latency, which is the total time elapsed between job submission and obtaining the results. Table 2 shows the latency for various bulk sizes of the workflow when submitted on three machines as against only one.

Table 2. Latency of workflows submitted to one machine and three machine grid.

No. of WorkFlows	Time Taken on 1 Machine	Time Taken on 3 Machines
10	3 hours 20 minutes	1 hour 28 minutes
20	6 hours 40 minutes	2 hours 5 minutes
25	8 hours 20 minutes	2 hours 33 minutes

3.3. Benefits of the Workflow

Workflow development and management enable faster execution of biological research activities such as functional annotation of genes or proteins and identifying homologous sequences. The ability to automate and accelerate the process without loss of quality in scientific output is the key benefit of this application. Considering the number of tasks whose execution is to be coordinated and completed, it is nearly impossible without the automation afforded by workflow management. The build, execution and reporting capabilities of the monolithic workflow manager can support large number of tasks over a distributed environment involving multiple heterogeneous platforms and multiple laboratories. The current capabilities of the workflow manager will be enhanced with increasing scientific challenges and demanding technological requirements in the near future.

The following are the benefits enjoyed by the clients once the solution is deployed: (i) Workflow can be created and saved for future use or re-engineer a new variant, (ii) Automated data conversions between components of the workflow, (iii) Reusable life science informatics application components, (iv) Promote collaboration activity, (v) Resource utilization by efficient task distribution and re-routing, (vi) Enhance experimental and computational research efficiencies.

4. CONCLUSIONS

Custom-built computational workflows builds, executes, manages and reports complex compute intensive workflows. Such workflows can support specific functional areas of discovery research e.g., gene identification, proteomics, compound screening, toxicology studies and pharmacogenetics. The workflow solution described in the paper (called DRWG) enables collaboration among various specialists, automates a pipeline of compute or data intensive work, and optimizes resource utilization. It manages repeated high throughput tasks, supports heterogeneous platform and reduces the cycle time of the process and together enhance productivity of scientific research.

ACKNOWLEDGMENTS

The authors are grateful to Shubhashis Sengupta, Sandeep Raju, Sanjay Martis and Deependra Moitra for discussion and help.

REFERENCES

1. Bernal A, Ear U, and Kyrpides, N. Genomes OnLine Database (GOLD): a monitor of genome projects world-wide. *Nucleic Acids Res* 2001; **29**: 126-127.
2. http://fightaidsathome.scripps.edu/index.html
3. Altschul SF, Gish W, Miller W, Myers EW, and Lipman DJ. Basic local alignment search tool. *J. Mol. Biol.* 1990; **215**: 403-410.
4. Thompson JD, Higgins DG, and Gibson TJ. CLUSTAL W: Improving the sensitivity of progressive multiple sequence alignment through sequence weighting, position-specific gap penalties and weight matrix choice. *Nucleic Acids Res.* 1994 Nov 11; **22**: 4673-80.
5. Durbin R, Eddy S, Krogh A, and Mitchison G. Biological Sequence Analysis: Probabilistic Models of Proteins and Nucleic Acids. Cambridge University Press, 1998.
6. Foster I, Kesselman C, and Tuecke S. The Anatomy of the Grid: Enabling Scalable Virtual Organizations. *International J. Supercomputer Applications* 2001; **15**(3): 327-344.
7. Hieter P and Boguski M. Functional genomics: it's all how you read it. *Science* 1997; **278** (5338):601-602.
8. Adams MD, Kelley JM, Gocayne JD, Dubnick M, Polymeropoulos MH, Xiao H, Merril CR, Wu A, Olde B, Moreno RF, et al. Complementary DNA sequencing: expressed sequence tags and human genome project. *Science* 1991 Jun 21; **252** (5013):1651-6.
9. Benson DA, Karsch-Mizrachi I, Lipman DJ, Ostell J, and Wheeler DL. GenBank: update. *Nucleic Acids Res.* 2004 Jan 1; **32** (Database issue):D23-26.

10. Pruitt KD, Tatusova T, and Maglott DR. NCBI Reference Sequence (RefSeq): a curated non-redundant sequence database of genomes, transcripts and proteins. *Nucleic Acids Res.* 2005; **33** (1):D501-D504.
11. Apweiler R, Bairoch A, and Wu CH. Protein sequence databases. *Curr. Opin. Chem. Biol.* 2004; **8**:76-80.
12. Gor K, Dheepak RA, Ali S, Alves L, Arurkar N, Gupta I, Chakrabarti A, Sharma A, and Sengupta S. Scalable enterprise level workflow and infrastructure management in a grid computing environment. *CCGrid*, May 2005.
13. Apache XML Beans, xmlbeans.apache.com
14. Tannenbaum T, Wright D, Miller K, and Livny M. Condor - A Distributed Job Scheduler. The MIT Press, 2002.

MOLWORKS+G: INTEGRATED PLATFORM FOR THE ACCELERATION OF MOLECULAR DESIGN BY GRID COMPUTING

FUMIKAZU KONISHI

RIKEN GSC, Bioinformatics Group, 1-7-22 Suehiro
Tsurumi, Yokohama, Kanagawa 230-0045, Japan
Email: fumikazu@gsc.riken.jp

TORU YAGI

BestSystems Inc., 1-1-1 Umezono, Tsukuba, Ibaraki
305-8568, Japan
Email: yagi@bestsystem.co.jp

AKIHIKO KONAGAYA

RIKEN GSC, Bioinformatics Group, 1-7-22 Suehiro
Tsurumi, Yokohama, Kanagawa 230-0045, Japan
Email: konagaya@gsc.riken.jp

In this paper, we present an integrated environment for a molecular design platform MolWorks+G based on Globus Toolkits. MolWorks+G allows you to connect computational resources as Grid Service, and simplifies an allocation of resources. As a key distinguishing feature, MolWorks+G supports a molecular builder and pre/post interface to quantum chemical calculation software, such as Q-Chem, Gaussian, MOPAC and GAMESS. MolWorks+G allows you to model a molecule structure by embedded Z-matrix editor for application data portability. MolWorks+G also provides an estimation of molecular properties which are implemented by the Joback method for thermodynamic properties, PVT diagram and so on. Thus, MolWorks+G is able to incorporate a significant application. We had developed MolWorks+G for computer aided material designs with the Grid Computing. MolWorks+G has allowed molecular science researchers to build modeling, to calculate large design, and to estimate high accurate property on Grid infrastructures.

1. INTRODUCTION

In silico drug design takes a computational challenge of molecular dynamics (MD) simulations for protein science brought about by the interactions among chemical compounds, genes, and proteins. The target simulations require mass computations will likely demonstrate significant result in the post-genomic era. In this direction, a peta FLOPS computer will be available for time-consuming and expensive applications such as MD simulation. Similarly, a virtual screening for target compound validation will play a key role in drug design for novel drug discovery.

A chemical compound database such as the Cambridge Structural Database[1] (CDC) registers 325,000 entries of the crystal structure information of the organic compounds and the organo-metallic compounds. It is used for a docking simulation by using MD approaches. However, there is a realistic issue about the preparation for the force field parameters which are calculated from each compound by quantum chemical calculation programs such as Gaussian[2] or Q-Chem[3]. Calculating a force field for each target compound for drug discovery needs a validation process of several candidates' structures by a chemist and/or pharmacologist. These researchers manage the validation task by sharing the load of force field calculations against limited computational resources.

In this paper, we present an integrated environment for a molecular design platform MolWorks+G based on Globus Toolkits[4]. MolWorks+G is an improvement of MolWorks[5,6] by taking advantage of the Grid environment. MolWorks+G allows you to connect computational resources as Grid Service, and simplify an allocation of resources.

2. FEATURES

2.1. Molecule Modeling

MolWorks+G provides a function for importing files such as XYZ format files (*.xyz), Mol files, Mol2 files and protein databank files (*.pdb). The XYZ format is a very simple scheme: The first line contains the number of atoms, the second line a is comment, and the remaining lines contain one atom each, with four entries: first the element symbol, and then the coordinates x, y, and z. In importing these file format, MolWorks+G allows you to display a molecular geometry for building the alternative structure easily. Currently, you can draw the molecule structure within "Molecule Window and Molecule Editor". MolWorks+G also provides "Optimizer" for relaxing the structure. It is easy to modify element types and bond orders with a selection of each atoms and bonds. MolWorks+G has the capability to display the molecular structures with wire, and ball & stick. And MolWorks+G also supports editing of the Cartesian coordinates and Z-Matrix. The latter defines connectivity between atoms in a molecule. The parameters required are distances, angles and dihedral angles.

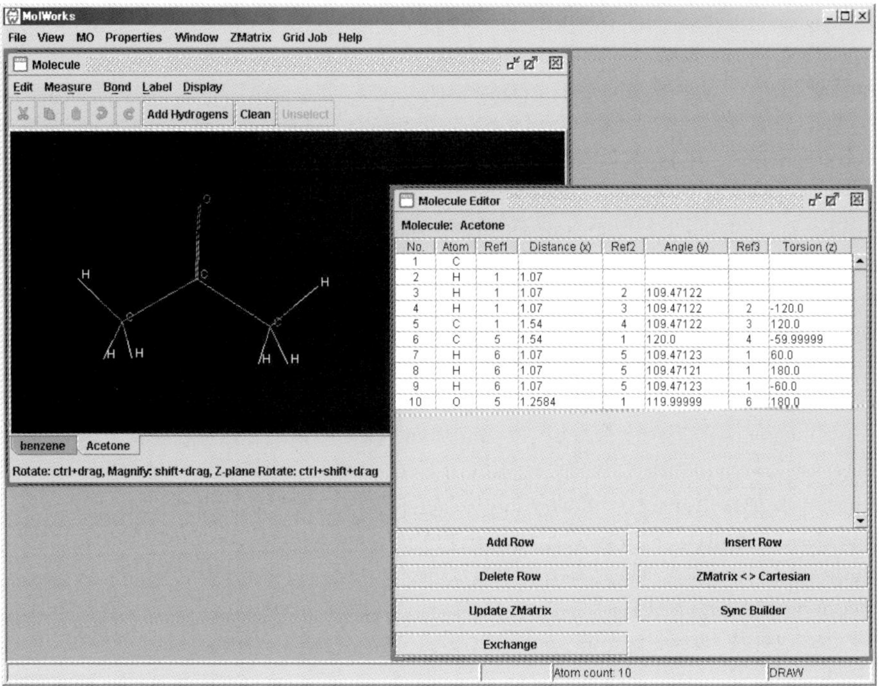

Figure 1. A snapshot of Molecule Window and Molecule Editor

2.2. Properties Estimation

MolWorks+G has implemented a property estimation features in modular architecture. The Joback method is a famous property estimation schemes by the group contribution method and is widely used[7]. In MolWorks+G, the Joback module allows you to estimate the properties of boiling point, melting point, critical temperature, critical pressure and critical volume, and to apply this capability for not only pure component, but also mixtures to get Pressure Volume Temperature (PVT) diagram.

Since MolWorks+G is made of the combination of a module group, it is easy to add a new function.

2.3. Pre-Post Processors

MolWorks+G has an embedded GUI for Q-Chem, Gaussian, GAMESS[8,] and MOPAC[9] on the MO Window, which consists of four tabs for each program. Each tab has several attributes such as calc. type, calc. level, basis set, polarization/diffuse, geometry and charge. MolWorks+G allows you to migrate attribute parameters for well-known quantum mechanics programs, like Q-Chem, MOPAC, GAMESS and Gaussian. For example, MolWorks+G can convert a Gaussian input parameters into a Q-Chem input parameters and create an input data file for Q-Chem. You can select the keywords and options for calculation from the menu in this GUI of quantum mechanics calculation assistant. In addition, MolWorks+G includes a CNDO/2[10] calculation engine and can display the MO (Molecular Orbital) within the Builder Panel. After you save the input file for performing a calculation, it is simple to submit those calculations for execution on a PC or high performance machines.

MolWorks+G can easily bridge a resource of Grid computing. First, MolWorks+G creates a proxy certification of Grid Security Infrastructure (GSI) for application users, and creates a job file which includes the information of specified host, job command file and application input file. Then MolWorks+G submits a job file by Globus Resource Allocation Manager (GRAM) client such as Java CoG kits[11], control the user jobs, and transfers the input file by GridFTP, monitors job status by using GRAM API periodically, and collects output files by using GridFTP when the job is completed.

MolWorks+G also bridges the gaps between the Grid environment and applications. A user can use the applications in the Grid environment by using only the GUI of MolWorks+G.

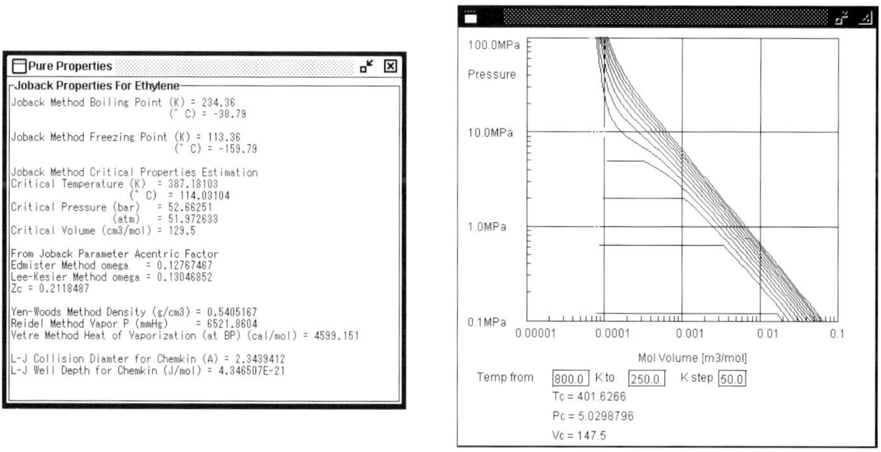

Figure 2. Snapshots of property window and PVT graph view

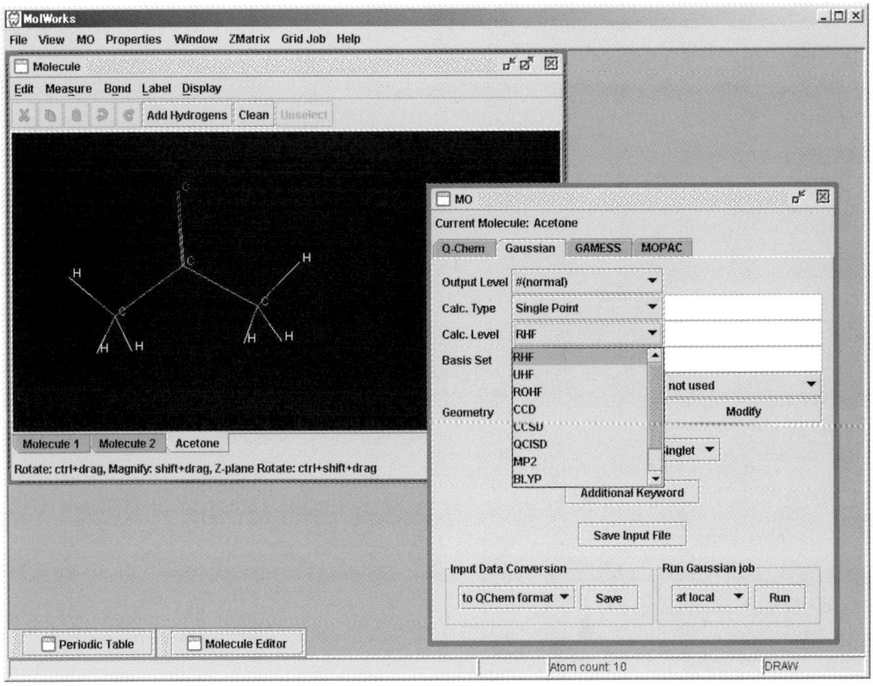

Figure 3. A snapshot of MO editor

3. ARCHITECTURES

In this section, we present the architecture and implementation of MolWorks+G. Figure 4 summarizes the architecture stack of MolWorks+G.

MolWorks+G has a modular architecture that allows a function builder to simply plug in components into the MolWorks+G. We have designed MolWorks+G as follows: independence of hardware, supporting Grid computing, and building the structure of molecules. To achieve hardware independence, MolWorks+G is written in the Java language. Since MolWorks+G has minimum hardware dependency, a user can utilize many platforms using the Java Run-time Environment.

MolWorks+G divides the functions of job control into two pieces, a meta-scheduler server and front-end client. The meta-scheduler server holds information about the parameters of quantum mechanics programs. The front-end client has a role of the interpreter between a user and pre-post processors, and a user can make a decision for submitting the job systematically. Communication to the

MolWorks+G server is implemented by XML data for description of job information. Two functional pieces can start on a same box or be separate physically. The meta- scheduler server periodically examines a remote job status by remote scheduler GRAM API, and can be tracking the current status of submitted user job.

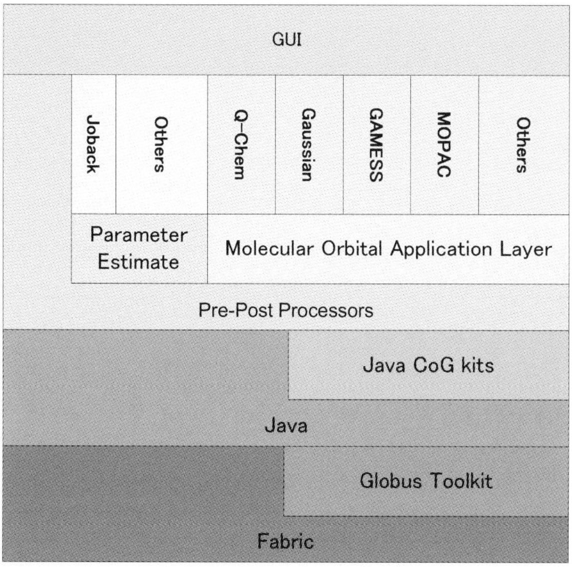

Figure 4. The architecture layer

4. CONCLUSIONS AND FUTURE WORK

We have developed an integrated application environment (MolWorks+G) for a molecular design platform. This is expected to assist molecular science researchers to build models, calculate large design, and estimate high accurate property on Grid infrastructures. For future work, we will incorporate a high availability meta-scheduler in MolWorks+G to improve system robustness and implement a function of high accurate and widely applicable property estimations based on a neural network method.

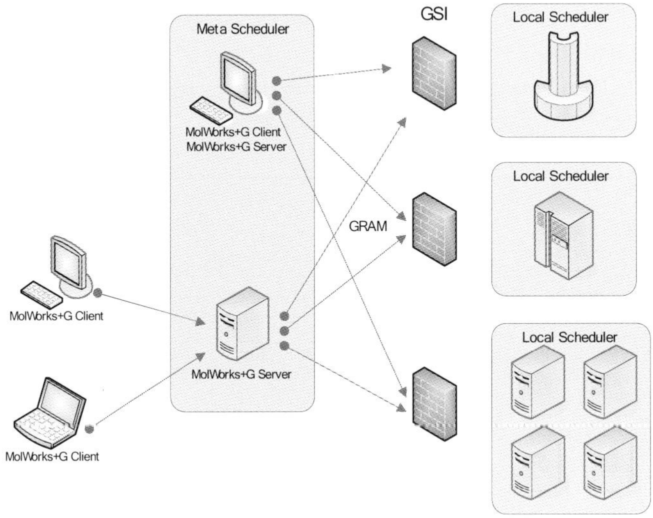

Figure 5. MolWorks+G System Architecture Overview

ACKNOWLEDGMENTS

We gratefully acknowledge the contributions of Katsuya Nishi (BestSystems, Inc.), Yusuke Hamano, and our co-workers. This work is supported in part by RIKEN Genomic Sciences Center and OBIGrid Project.

REFERENCES

1. The Cambridge Structural Database: a quarter of a million crystal structures and rising F. H. Allen, Acta Crystallogr., B58, 380-388, 2002
2. Gaussian - http://www.gaussian.com/
3. Q-Chem - http://www.q-chem.com/
4. Globus Toolkit - http://www.globus.org/
5. Tajima, S., Nagashima, U., and Hosoya, H., Journal of Computational Chemistry Japan, 2002, 1, 103.
6. MolWorks – http://www.molworks.com
7. Joback, K. G., S.M. Thesis in Chemical engineering, Massachusetts Institute of Technology, Cambridge, Massachusetts, **1984**.
8. GAMESS - http://www.msg.ameslab.gov/GAMESS
9. Stewart, J.J.P., *J. Comput. Chem.*, **1989**, *10*, 209, 221.

10. Pople, J. A. and Beveridge, D. L., "Approximate Molecular Orbital Theory", McGraw-Hill, New York, **1970**.
11. Java CoG kit – http://www.cogkit.org/

PROTEOME ANALYSIS USING IGAP IN GFARM*

WILFRED W. LI†, PETER W. ARZBERGER‡

National Biomedical Computation Resource, Life Science Initiatives, University of California, San Diego, 9500 Gilman Drive, La Jolla, CA 92093-0505
Email: {wilfred,parzberg}@ucsd.edu

CHANG LIM YEO, LARRY ANG

Bioinformatics Institute, 30 Biopolis Street
#07-01, Matrix, Singapore 138671
Email: {yeocl, larry}@bii.a-star.edu.sg

OSAMU TATEBE, SATOSHI SEKIGUCHI

Grid Technology Research Center, AIST, 1-1-1 Umezono, Tsukuba, Ibaraki, Japan
Email: {o.tatebe, s.sekiguchi}@aist.go.jp

KARPJOO JEONG

College of Information and Communication, and Bio/Molecular Informatics Center, Konkuk University, 1 Hwaynag-dong Gwangjin-gu, Seoul 143-701, South Korea
Email: jeongk@konkuk.ac.kr

SUNTAE HWANG

School of Computer Science, Kookmin University, Jeongneung-dong, Songbuk-gu
Seoul 136-702, South Korea
Email: sthwang@kookmin.ac.kr

SUSUMU DATE

Department of Bioinformatic Engineering, School of Information Science & Technology, Osaka University, 1-3 Machikaneyama-cho, Toyonaka, Osaka 560-8531, Japan
Email: date@ais.cmc.osaka-u.ac.jp

* This work is funded by independent grants to the participants from respective governmental funding agencies.

† Work partially supported by the PRAGMA grant INT-0314015 of the National Science Foundation.

‡ Work partially supported by the NBCR grant P41 RR 008605 of the National Center for Research Resources, NIH.

JAE-HYUCK KWAK

Grid Technology Research Department, Supercomputing Center, KISTI, Eoeun-Dong 52, Yuseong-Gu, Daejeon 305-806, South Korea

Email: jhkwak@kisti.re.kr

The Integrative Genome Annotation Pipeline (iGAP) is a suite of bioinformatics software developed for annotation of protein structure and functions. It was previously deployed on the grid using APST from UCSD, Grid Monitor from BII, and GridSpeed from TiTech. The distributed nature of the grid necessitates the development of a global parallel file system (GPFS). Gfarm™, developed by AIST in collaboration with KEK, University of Tokyo, and TiTech, is such a GPFS. Gfarm is originally designed as a high performance peta-scale, parallel data transfer and computation platform. Version 1.0.4 of Gfarm enables the execution of existing applications within the Gfarm virtual file system with no modifications required. This new development, important for many bioinformatics applications, allows the successful execution of iGAP in Gfarm using distributed compute and storage resources transparently. We report our experience of running iGAP in Gfarm to analyze the complete proteome of the bacteria *Burkholderia mallei*, a known Category B biothreat agent, on resources distributed in 7 universities in 4 countries of the Pacific Rim. The applications compiled for different compute platforms/architectures are installed in Gfarm in the same apparent global path. The correct architecture is selected transparently on the fly. The data libraries required for the calculations are loaded in one compute resource, and automatically replicated to other compute resources on demand. Because the data is located locally on each compute node with a public IP address, the application executes efficiently. In addition, the "gfrun" remote execution command, which schedules jobs using file affinity scheduling, proves to be efficient as well. The bottleneck, however, proved to be the registration of thousands of files at the Gfarm metadata server. Version 1.1.1 of Gfarm released in April 2005 significantly improves the file registration performance. We have performed benchmark studies using iGAP, and the current Gfarm grid environment performs well compared to the cluster environment. Running iGAP inside Gfarm not only offered a real life science application suite using a new technology not available before to life sciences applications, but also provided new sparks and momentum for Gfarm software development and code hardening, as new features planned in v2 of Gfarm will provide further enhancement. The "install once and run anywhere" feature of Gfarm is a major step towards lowering the cost of entry for the routine use of the grid in bioinformatics.

1. INTEGRATIVE GENOME ANNOTATION PIPELINE (IGAP)

1.1. Grid Enabling Bioinformatics Applications

The international genome sequencing effort has steadily produced a large number of complete genomes. Since 1995, more than 180 complete, and over 1000 partial proteomes have been made publicly available. Progress is being made towards high quality proteome annotation through a combination of high throughput computation as well as manual curation. Increasingly, the new knowledge is translated into tangible diagnostic and remedial procedures for the benefits of public health and education. Grid technology promises to meet the increasing demand in large scale computation and simulation in the fields of computational biology, chemistry and bioinformatics. The integrative Genome Annotation Pipeline, iGAP[1], provides functional annotation using known protein structural information. The computational requirement of iGAP and our initial experience in using AppLeS Parameter Sweep Template (APST)[2] to deploy it on the grid has been previously described[3]. In addition, a prototype Bioinformatics Workflow Management System (BWMS) integrates with APST to facilitate the multi-genome annotation effort[4]. To facilitate the interactive use of the pipeline and the monitoring of the annotation progress, a GridSpeed based application portal, and GridMonitor Portal was also developed[5].

While the previous system provides many rich features and works well for the dedicated workflow, it requires detailed knowledge to operate and requires significant effort to generalize to other applications. Ease of use is fundamental for the widespread adoption of grid technologies in life sciences, where application scientists do not have the time to keep up with the ever changing grid computation standards and models. As Moore's law continues to hold true, commodity computing clusters are becoming a reality on university campuses across the globe. While this trend is fundamental to the maturation of the grid, it also poses a new challenge. Many users prefer to run bioinformatics applications within a cluster environment, which is the most reliable production environment to date, despite recent advances in grid middleware technology.

One problem we have experienced in both the cluster and grid environments is the limitation of file I/O. In a cluster environment, if an application generates a lot of intermediary and output files, the load on the NFS server may become quite high, as the number of compute nodes increases, and I/O becomes a rate limiting step. Code modification is required to move the file I/O to local disks on compute nodes. Additional code is required to reliably transfer the end results from the compute nodes back to the NFS server. Even with a dedicated cluster where local disk may be used for data storage, it becomes difficult to find the results on various compute nodes if one wishes to revisit the data after a certain period of time. In a grid environment, network latency, bandwidth shortage, and overhead in file transfer are often prohibitive to the effective grid deployment of legacy applications which produce many output files with sizes ranging from tens of kilobytes to tens of megabytes. Gfarm™ provides a scalable and transparent solution, which enables effective use of not only the distributed computing power, but also the disk storage space, through a familiar virtual file system view (see Figure 1).

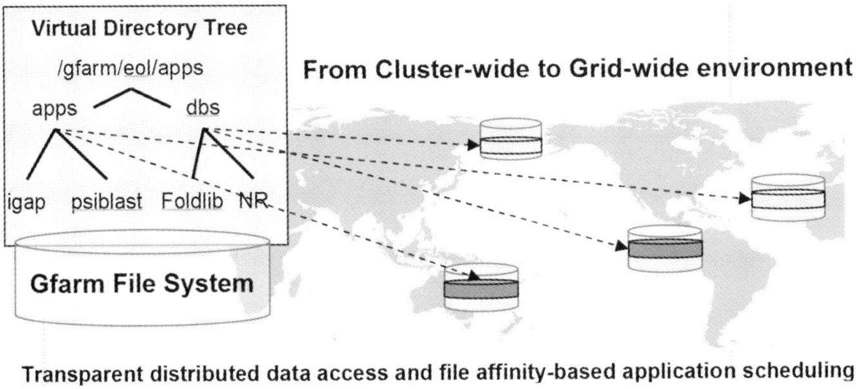

Figure 1. Gfarm file system enables centralized distribution of biological databases and applications.

2. GFARM VIRTUAL FILESYSTEM

2.1. Gfarm

The Grid Datafarm[6] architecture is designed for global petabyte scale data-intensive computing, which provides a Grid file system with file replica management (Gfarm file system), and parallel and distributed data processing support for a set of files (*Gfarm file*). It provides scalable I/O bandwidth, and scalable parallel processing to exploit local I/O in a grid of clusters. The data is, physically replicated and dispersed among cluster nodes across administrative domains, where it can be accessed transparently from file replica locations via POSIX file I/O interface by data analysis tools.

The most time-consuming but also the most typical task in various fields, such as astronomy, high energy physics, space exploration and human genome analysis, is to process and produce sets of files in embarrassingly parallel fashion. Such a process can be typically performed independently on every file in parallel, or at least have good locality. Gfarm supports high-performance distributed and parallel computing for such a process by introducing a *Gfarm file*, a new *file-affinity* process scheduling based on file locations, and new parallel file access semantics. An arbitrary group of files possibly dispersed across administrative domains can be managed as a single Gfarm file. File affinity scheduling and file view feature naturally drives the *owner computes* strategy, or *move the computation to data* approach. This is the key distinction of Gfarm over other distributed file systems, where the data will be moved to computation by default.

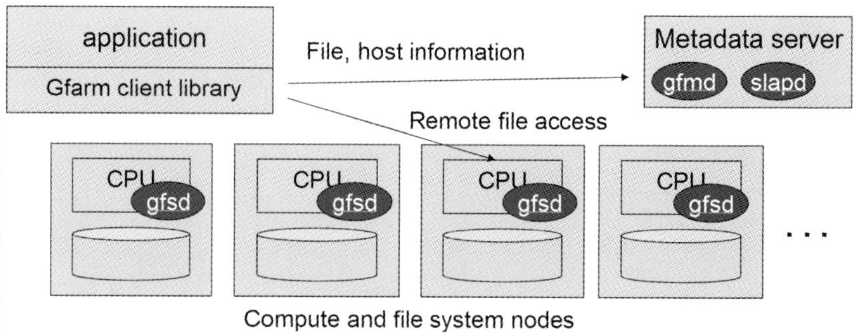

Figure 2. Gfarm file system architecture. OpenLDAP is used for keeping track of meta data. Gfmd and slapd do not have to reside on the same server. A metadata server may also be a file system node. The data and compute nodes are one and the same, thereby reducing unnecessary data transfer.

While Gfarm files may be accessed by a parallel I/O APIs (Gfarm APIs), it requires a series of modifications, first of application source code, then from subsequent API changes, not to mention possible newly introduced bugs. Therefore, as of version 1.0.4 released in November 2004, Gfarm provides a system call hooking library which enables existing applications to run in Gfarm without code modification. It traps system calls for file I/O to investigate whether the specified operation is for a Gfarm file system or not. If it is for a Gfarm file system, it calls appropriate Gfarm APIs. This means that many applications in the life sciences may be able to leverage the grid computing power in Gfarm. We have set out to collaborate under PRAGMA to establish such a prototype grid environment using iGAP. Through rounds of software debugging and testing, version 1.1.1 of Gfarm released in April 2005 provides such a stable and efficient grid environment for iGAP, with no changes in iGAP required.

3. PROTEOME ANALYSIS IN GFARM

3.1. Building and Using a Transparent Gfarm File System

With participants from 6 institutes during SC'04, we have set up a Gfarm testbed for iGAP, http://datafarm.apgrid.org/testbed/igap/ (see Table 1). The goal was to use iGAP to analyze the complete proteome of the bacteria *Burkholderia mallei*, a known Category B biothreat agent, on internationally distributed resources, by moving the applications to the data, without the data transfer overhead. The applications compiled for different compute platforms/architectures are installed in Gfarm in the same apparent global path. The correct architecture is selected transparently on the fly. The data libraries required for the calculations are loaded in one compute resource, and automatically replicated to other compute resources on demand.

In Gfarm VFS, a typical command would be:

```
[wilfred@rocks-32 igap]$ gfrun sh \
/gfarm/wilfred/eol/apps/igap/igap_1.22.sh -r \
/gfarm/wilfred/.igap/burma_tigr_110904 -psiblast -matrix_single \
-sequence 1833 -X
```

Compared to the same command in the NFS, the only difference is the replacement of /gfarm/wilfred with /home/wilfred. In the mean time, the user now has access to a grid of clusters with significant computing power and storage space (Table 1).

Table 1. A 52 GB RAM, 202 GFLOP, and 2.5 TB peak capacity Gfarm testbed for iGAP. N is the number of nodes. P is the number of processors per node.

	N	P	Type	Proc Speed (GHz)	Memory (GB)	Hard Disk (GB)	OS	Switch (Gbps)
Kookmin	8	1	AMD Athlon	1	0.256	40	RH 7.3	1
Konkuk	8	1	Intel P4	3	1	40	RH 7.3	1
KISTI	8	1	Intel P4	2.4	1	80	RH 7.3	1
UCSD	5	2	Intel Xeon	3.06	2	40	Rocks 3.2	1
BII	4	1	Intel P3	1.3	1	30	RH 8.0	.008
Osaka	10	1	Intel P3	1.4	1	70	RH 7.2	1
Total	43	48		202	52.048	2500		

While the theoretical peak performance is 202 GFLOPS and 2.5 TB of disk storage, prior disk usage reduced the available space to 1 TB. Using thput-gfpio, part of the file system node (fsnode) package in Gfarm, we observed an overall 1 GB/sec disk I/O performance.

Overall, the building of the testbed from source RPM or binary RPM are straight forward and well documented. In order to enable transparent access to the virtual file system without code modification, a glibc-not-hidden package is required. A dedicated metadata server is recommended, where all the information about the files and applications are kept (See Figure 2).

Gfarm currently uses LDAP server for tracking directory, file and file replica information. We have found that openldap-2.1.30, with BerkeleyDB, performs much better than the openldap-2.0.27, which uses LDBM (GNU dbm). Additionally, increasing the amount of memory cache significantly improves the performance of the LDAP server. The cache size was set to 2 GB, using a 2.8 GHz, dual 64-bit AMD processor with 3 GB of RAM, and 70 GB of local disk. Configuration and performance tuning of Gfarm are discussed in subsequent sections.

3.2. Performance Tuning

3.2.1. *Comparison to NFS*

NFS file system is often used in a clustered environment. However, when file I/O is heavy, the performance does not scale well and could be very poor. Since Gfarm takes advantage of local disks on compute nodes and turns them into storage nodes as well, the potential for scalability and high performance is very good.

The main new feature, in Gfarm v1.1.1, realizing a significant performance improvement, is as follows: Gfarm Agent (GA) caches pathname and timestamp of directories and files. In the authors' experience, after the initial caching delay of a few seconds, directory listing occurs on the scale of milliseconds, even when there are thousands of files in a directory.

Table 2. Performance comparison using iGAP-WUBLAST measured in seconds. Each measurement is the average of two separate experiments. Using Gfarm Agent and increasing LDAP memory cache significantly increases the performance of WUBLAST of a 100 sequence dataset.

iGAP-WUBLAST performance on 100 sequences (s)	Local NFS	Remote Gfarm			
		Gfarm Agent			
				LDAP Memory	
real	46.226	311.721	212.086	156.675	188.448
user	141.385	187.475	218.84	159.9	0.065
sys	5.565	11.475	10.69	10.775	0.025

Using iGAP executing WUBLAST as an example, its performance in Gfarm without, with Gfarm Agent, and the latter with LDAP memory cache optimization of Gfarm is compared to that in NFS. As shown in Table 2, the user time for NFS and Gfarm are different by about 14% within the same node (141.385 vs. 159.9) and about 33% when iGAP-WUBLAST is executed on a remote node using gfrun (141.385 vs. 188.448). The real time difference of about 3.4 fold is because WUBLAST takes advantage of multiple threads in native NFS, whereas multithreading is not supported in Gfarm currently. In the case of iGAP-PSI-BLAST, where multithreading is disabled by a user supplied option, the difference in real time performance is down to about 16% (see Table 3).

Table 3. The performance of iGAP-PSI-BLAST measured in seconds. Each value is the average of two separate experiments.

iGAP-PSI-BLAST performance on 2 sequences	Local	remote			
	NFS	Gfarm			
		Gfarm Agent			
		Pre-staged		replication	
Real	130.882	233.032	345.41	218.103	155.409

Also notable in Table 3 is the effect of the on-demand replication feature of Gfarm. When the required library for PSI-BLAST is not pre -staged or -replicated, the difference could be close to 66% (130.882 vs. 218.103). Without Gfarm Agent, the different may be up to 163% (130.882 vs. 345.41). After the initial replication, a second run with all the prerequisite files lead to a 2.2 fold increase in PSI-BLAST performance in Gfarm.

3.2.2. *Job scheduling*

Gfarm uses file affinity scheduling, where the application is executed where the required data is available. It also takes into account of the load average on the compute nodes. While load average may be used for scheduling, sometimes network latency may cause the information to be dated, and cause many jobs to be scheduled to a node. While job scheduling in Gfarm is an area of active collaboration under PRAGMA[7], we are currently using a simple FIFO scheduling approach. Basically each node is scheduled the same number of jobs as the number of processors.

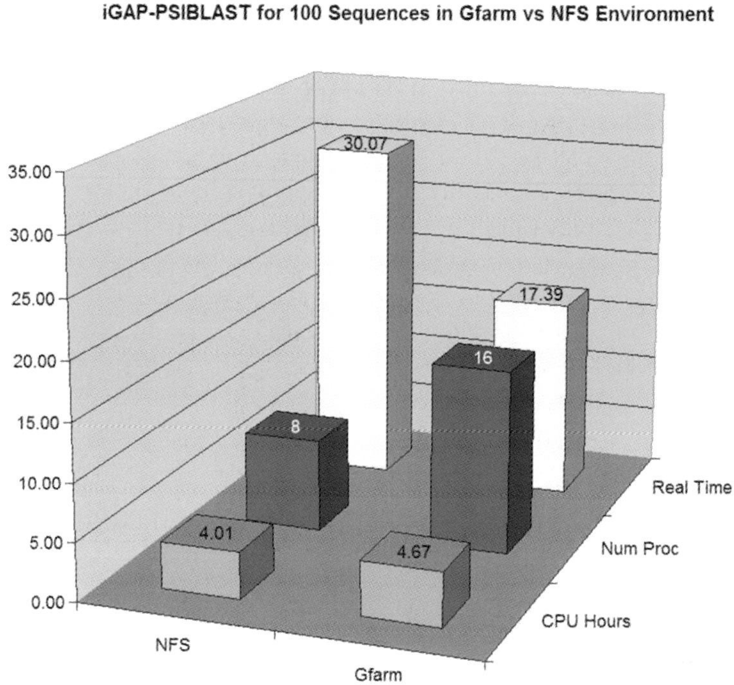

Figure 3. Average CPU hours required to process 100 sequences using iGAP-PSI-BLAST in Gfarm and NFS file systems. The NFS is a shared file system for a 4 node cluster. The Gfarm is two separate 4 node clusters. 4 cpu hours is required on NFS vs. 4.67 CPU.Hours on the Gfarm.

Figure 3 compares the performance of a NFS based cluster environment, against a Gfarm grid environment consisting of two separate clusters. Using FIFO, we are able to achieve about 86% of the performance, in terms of CPU hours, of an NFS file system for a 100-sequence data set using PSI-BLAST. While Gfarm only achieved a 1.7 fold speedup using twice as many processors, the 14% difference does provide additional incentive to further optimize the performance of Gfarm. Previously we have used APST to schedule jobs to various grid resources. But we have often encountered a bottleneck with transferring the files back from remote resources. With Gfarm, the files are not transferred back, but only their locations are recorded. Should there be a need to visit these intermediary files, it is very easy to retrieve them, just like in a regular Unix Shell. This reduction in file transfer overhead is one appealing aspect of using Gfarm. The 0.67 hour over NFS could easily be compensated from a practical perspective when the time consumed to manage files in different clusters is taken into account.

3.3. Network Communication Tuning

Because distributed file systems need to communicate the metadata, as well as moving data over potentially congested network. It's necessary to use best available network if possible. Gfarm has been shown previously to achieve very high performance on transfer of large data set, and won the Distributed Infrastructure award during SC'03. Using Iperf[8], we have also tuned the TCP/IP buffer size, as well as the mem_max allowed on compute nodes. These significantly improve the file replication performance. In our experiment, we found it necessary to increase the TCP/IP buffer size by setting the following parameters:

1. net.core.r(w)mem_max=1048576 -- /etc/sysctl.conf
2. net.core.r(w)mem_default=524288
3. sockopt snd(rcv)buf=307200 -- /etc/gfarm.conf
4. netparam parallel_streams=4

After tuning these parameters, the file transfer speed reaches 2.5 Mbps per second between BII and SDSC. While this is still only about 10% of the performance of local disk I/O speed, it is reasonable given the 8Mbps local network switch for the BII workstations. At this rate, an 800 MB NR file takes about 30 min to replicate. For practical reasons, we decided to gfrep to pre-stage these files at BII instead of using on demand replication. On the other hand, between Osaka and SDSC, transfer of the same file only took 6.5 min, achieving 18.1 Mbps. However, a file of 107 MB only achieved 9.3 Mbps. This, along with our observations using smaller files, suggests that Gfarm is not yet optimized for small file transfers.

3.4. Metadata Server Tuning and Gfarm System Status Monitoring

The LDAP server is a critical component in Gfarm performance. Although the inclusion of Gfarm Agent significantly reduces the load on the LDAP server, its performance, stability and backup are critical to Gfarm performance and reliability.

We have found the following parameters for BerkeleyDB to be important in DB_CONFIG:

(a) set_cachesize 0 536870912 2
(b) set_lg_bsize 2097152
(c) set_flags DB_TXN_NOSYNC

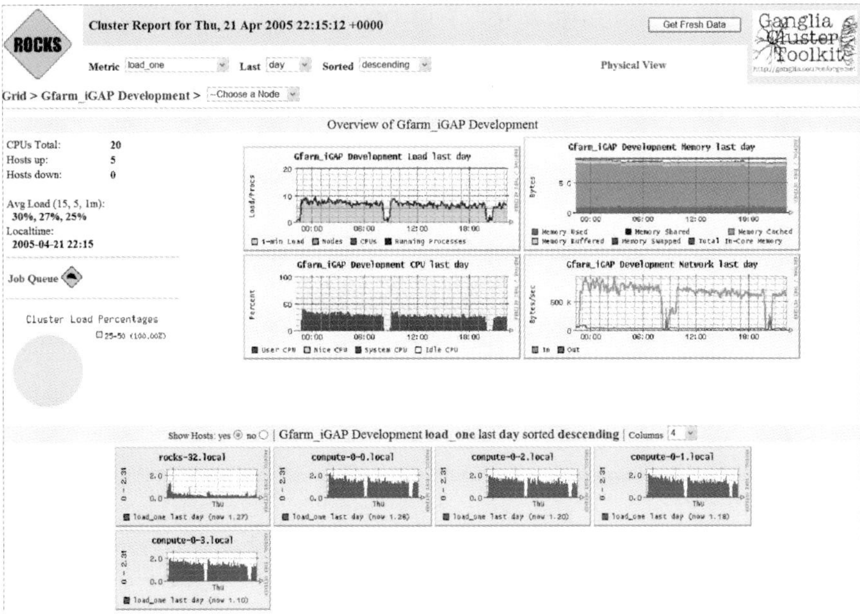

Figure 4. Monitoring a Gfarm file system through the Ganglia interface. A PSI-BLAST run of 2000 sequences is accomplished in 9 hours, comparable to a NFS run using SGE.

For performance reasons, LDAP logging is disabled. Access to the LDAP database server is restricted by IP addresses for security reasons. A cron job is set up to back the LDAP database, while gfsplck and gfsck are used to verify the integrity of the metadata on a regular basis.

The status of Gfarm file system can be monitored easily using the Ganglia cluster toolkit, which comes standard with the Rocks distribution. As shown in Figure 4, the network, memory, disk usage, and cpu load are graphically depicted. The monitoring is important because if network is down, or interrupted, or if a particular host's disk is full, Gfarm performance will be affected.

4. DISCUSSION

Collaborating under PRAGMA, we have established a prototype grid environment for scientific applications without code modification. The collaboration has been productive in terms of pushing the stability and performance of Gfarm. What is the main advantage of Gfarm? It reduces the cost barrier of using the grid for application scientists. Because there is no code modification required, the application scientist can focus on the science while leveraging grid resources. When computational resource is scarce, access to Gfarm grid environment would provide much expanded computational power, as well as storage resources, as shown in Table 1.

Currently file access in Gfarm uses Unix file system protection. However, metadata is currently not protected. Thus, access to metadata is only restricted using Unix firewall, to compute nodes. User authentication is accomplished using shared key, or GSI authentication. The latter is more desirable, because of its session management and expiration policy, whereas the shared key method expires every 24 hours. Even though Gfarm supports data encryption and integrity by GSI, we do not use these features because they slow down the performance by up to 5 times.

Since we started experimenting with Gfarm v1.0.4, the performance and stability of Gfarm has improved significantly. With the release of v1.1.1, we have established a routine use grid environment in Gfarm. The genome of *Bur. mallei* has been completed, and a scientific analysis and comparison with its non-pathogenic neighbors are being conducted. The results will be reported separately.

Clearly if one wants to take full advantage of a particular system, it is necessary to access some low level API to reduce the overhead and improve performance. However, given the state of flux of grid technology, Gfarm provides a stable environment for practical use of the data and computational grid in the life sciences.

ACKNOWLEDGMENTS

We would like to thank Taehoon Kim of Konkuk, Gyuho Sim, Daeyoung Heo of Kookmin, Young-Chul Hwang of KISTI, Takuji Nakagawa of Osaka, Bernard Kian Meng Tan of BII, and Cindy Zheng of UCSD/SDSC for their assistance in setting up the Gfarm virtual file system.

REFERENCES

1. A comparative proteomics resource: proteins of Arabidopsis thaliana: Li, W. W.; Quinn, G. B.; Alexandrov, N. N.; Bourne, P. E.; Shindyalov, I. N., A comparative proteomics resource: proteins of Arabidopsis thaliana. *Genome Biol* **2003,** 4, (8), R51
2. Parameter sweeps on the Grid with APST: Casanova, H.; Berman, F., Parameter sweeps on the Grid with APST. In *Grid Computing: Making the Global Infrastructure a Reality*, Berman, F.; Fox, G. C.; Hey, A. J. G., Eds. Wiley Publishers, Inc.: West Sussex, 2003.

3. The Encyclopedia of Life Project: Grid Software and Deployment: Li, W. W.; Byrnes, R. W.; Hayes, J.; Birnbaum, A.; Reyes, V. M.; Shahab, A.; Mosley, C.; Pekurovsky, D.; Quinn, G. B.; Shindyalov, I. N.; Casanova, H.; Ang, L.; Berman, F.; Arzberger, P. W.; Miller, M. A.; Bourne, P. E., The Encyclopedia of Life Project: Grid Software and Deployment. *New Generation Computing* **2004,** In Press.
4. Grid Workflow Software for High-Throughput Proteome Annotation Pipeline: Birnbaum, A.; Hayes, J.; Li, W. W.; Miller, M. A.; Arzberger, P. W.; Bourne, P. E.; Casanova, H., Grid Workflow Software for High-Throughput Proteome Annotation Pipeline. *Lecture Notes In Computer Science* **2004,** In Press.
5. Grid Portal Interface for Interactive Use and Monitoring of High-Throughput Proteome Annotation: Shahab, A.; Chuon, D.; Suzumura, T.; Li, W. W.; Byrnes, R. W.; Tanaka, K.; Ang, L.; Matsuoka, S.; Bourne, P. E.; Miller, M. A.; Arzberger, P. W., Grid Portal Interface for Interactive Use and Monitoring of High-Throughput Proteome Annotation. *Lecture Notes In Computer Science* **2004,** In Press.
6. Grid Datafarm. http://datafarm.apgrid.org/software/#download
7. Implementing data aware scheduling on Gfarm by using LSF™ scheduler Plugin: Wei, X.; Li, W. W.; Tatebe, O.; Xu, G.; Hu, L.; Ju, J. In *Implementing data aware scheduling on Gfarm by using LSF™ scheduler Plugin*, International Symposium on Grid Computing and Applications, Las Vegas, NV, 2005; Las Vegas, NV, 2005; p In Press.
8. Iperf. http://dast.nlanr.net/Projects/Iperf/#download

GEMSTONE: GRID ENABLED MOLECULAR SCIENCE THROUGH ONLINE NETWORKED ENVIRONMENTS

KIM BALDRIDGE[A,B], KARAN BHATIA[A], BRENT STEARN[A], JERRY P. GREENBERG[A], STEPHEN MOCK[A], SRIRAM KRISHNAN[A], WIBKE SUDHOLT[B], ANNE BOWEN[A], CELINE AMOREIRA[B], YOHANN POTIER[B]

[a]San Diego Supercomputer Center, University of California, San Diego
9500 Gilman Drive, La Jolla, CA 92093-0505
[b]University of Zurich, 190 Winterthurerstrasse, Zurich, Switzerland, CH 8057

The life sciences have entered a new era defined by large multi-scale efforts conducted by interdisciplinary teams, with fast evolving technologies. Ramifications of these changes include accessibility of diverse applications as well as vast amounts of data, which need to be processed and turned into information and new knowledge. Accessibility via a multitude of clients, and ability to enable composition of data and applications in novel ways for facilitating innovation across an interdisciplinary group of scientists is most desirable. However, issues of diverse data formats and styles must be addressed to enable seamless interoperability. Adding Web service wrappers alleviates many problems because communication is by use of strongly typed data defined using XML schemas. Workflow tools can then mediate the flow of data between applications and compose them into meaningful scientific pipelines. This work describes the development of an integrated framework for accessing grid resources that supports scientific exploration, workflow capture and replay, and a dynamic services oriented architecture. The framework, Grid-Enabled Molecular Science through Online Networked Environments, GEMSTONE, provides researchers in the molecular sciences with a tool to discover remote grid application services and compose them as appropriate to the chemical and physical nature of the problem at hand. The initial set of applications services to date includes molecular quantum and classical chemistries together with supporting services for visualization, databases, auxiliary chemistry services, as well as documentation and educational materials.

1. INTRODUCTION

The fundamental goal in molecular modeling research is to understand how the interplay of structural, chemical and electrical signals gives rise to functions in biological systems. Studies conducted over the past 100 years have revealed the incredible complexity of biological systems. Experimental advances of the past decades additionally provide increasingly powerful arsenal for obtaining data, from the level of molecules to more complex macromolecular structure and beyond. However, addressing the consequences of molecular and structural specialization and variation across molecular scales through experimentation and/or computation is still challenging for a variety of reasons, not the least of which is the complexity of the computation involved and establishing a mapping between the computational data and experimental phenomenon.

Computational scientists have at their disposal several powerful simulation techniques which provide the means to explore the functional consequences of variation in structural features, i.e., shape, size, molecular constituents, and provide a necessary complement to existing experimental approaches. These models are, by nature, interdisciplinary in that they integrate information derived from multiple disciplines and across scale. Computational modeling is typically viewed in a hierarchical fashion, where one traverses from the molecular to the organelle scale. Although each modeling application tends to be optimized for events occurring at a single level, each is critically dependent on parameters passed from larger and smaller domains of function. For example, molecular properties such as kinetic parameters, diffusion constants, binding constants, channel locations and densities are required for physiological modeling on the micro-domain and cellular scale.

First principles, (*ab initio*) computational methods offer the most accurate approximations of the Schrödinger Equation and molecular Hamiltonians, from which molecular structure, mechanisms, properties, and dynamics, are determined. Such enormously powerful techniques have, to date, had little direct impact on the prediction of macroscopic properties of biological materials. Detailed quantum mechanical treatments of molecular and electronic structure can predict molecular geometry, follow the reaction paths of chemical transformations, predict electrostatic effects in a variety of environments, estimate pK_a shifts, and provide interpretations of spectroscopic probes of molecular environments.

As the molecular system size grows, the use of quantum chemical methodologies is either replaced or interfaced to less accurate, more affordable, methodologies, such as classical force field or electrostatic methodologies. The different applications, while related, each represent a different algorithmic methodology for understanding molecule structure and properties, and in many areas of research, these as well as other related applications may be accessed by scientists in one of many combinations. Innovative couplings between a wide range of different but related applications, including quantum chemical codes (e.g., GAMESS), classical codes (e.g., APBS and AMBER), docking applications, and hybrid modeling, is becoming an important aspect of computational modeling.

To achieve this goal, we are creating an integrated framework for accessing grid resources that supports scientific exploration, workflow capture and replay, and a dynamic services oriented architecture. The framework, called GEMSTONE [1] for 'grid enabled molecular science through online networked environments', provides researchers in the molecular sciences with a tool to discover remote grid application services and compose them as appropriate to the chemical and physical nature of the problem at hand. The initial set of application services focuses on bridging the domains of computational chemistry and molecular electrostatics, together with services for visualization, databases, and auxiliary chemistry services for applications in the Life Sciences.

2. BUILDING A DYNAMIC AND FLEXIBLE GRID ENVIRONMENT

A primary focus of GEMSTONE is on "composable grid application services architecture." There are three major components in this focus,

Component 1. Backend workflow middleware, InFORMNet
Component 2. Data type system
Component 3. User interface, GEMSTONE

Components 1 and 2 have been the focus of the early stages of our efforts, and provided sufficient experience and knowledge of middleware capabilities to pursue component 3. Our vision is the establishment of a truly usable and integrated environment for computational chemistry and biochemistry applications as an integration of the high-end capabilities of grid computing, the usability of web-based portals, and the flexibility of real time user-created workflows. While additional work on components 1 and 2 is ongoing, our most recent progress involves component 3, the user interface. As such, we will only briefly describe the first two components, referring to our previous publications in these areas, and concentrate on component 3 here.

Component 1: The first component, the workflow middleware, encompassed the early stages of development of this work, focusing on the design and development of the workflow system that handled INformation Flow & Operation Resource Management on the NET, called Informnet [2-4]. Informnet is a service-based workflow system that operates in conjunction with real-world scientific applications such as the GAMESS [5] quantum chemistry code, which is developed and maintained in this group. The core of the system is a workflow engine based on the Globus Toolkit 3 [6] and an XML schema-based workflow description language that facilitates late binding of tasks to computational resources by allowing multiple computational resources to be associated with a specific task in the workflow. The workflow engine grid service executes the tasks described by the workflow XML document while the client controls execution of the workflow and receives notifications of status, errors, and output from the workflow engine. We have also built a workflow composition tool that can be used as a client to the workflow engine. The workflow composition tool is a Java application that graphically constructs workflows for the user, building an XML document in the background. The workflow composition tool can then be used as a client to the workflow engine by sending the workflow description to the engine and controlling execution of the workflow. The service based workflow engine is capable of continuing execution of the workflow after the client disconnects. Many workflows run for long enough periods of time, such that the user cannot be expected to run the workflow program on their desktop workstation for the entire span of time that it takes for the workflow to execute. Informnet allows the client to disconnect from the engine and have the engine continue execution, then reconnect to the service later to regain control and collect status information. This workflow system, is currently functional, however, additional work in this phase is ongoing to support workflow publication and discovery, as well as brokering and fault tolerance, a straightforward extension of the initial efforts. We also are working towards automatic generation of workflows based on user interaction, which is more challenging, as properly generalizing from interactions of any particular researcher requires some experimentation.

Component 2: In the design of the workflow, we developed a domain-specific type system that describes the inputs to and the outputs from each task in the workflow graph. Correspondingly, the same type system is used to describe the inputs to the applications running on the computational resources. This enables strong type checking in the system and simplified output verification and data parsing. The type system we have developed is described by a set of XML schemas and instantiated as an XML document using our defined schemas. Our data type system is currently oriented towards the GAMESS quantum chemistry code and its auxiliary applications such as PLTORB3D [7], QMView [8, 9], and APBS [10], the first two of which have ongoing developments in this group. The schema contains the essential data that is output from a quantum chemistry code including atomic coordinates, molecular orbital data, and first and second derivatives of the energy with respect to the coordinates. As one of the authors of GAMESS is involved in this work, we have started the integration of these types directly and natively into the application codes themselves. This allows GAMESS to read and write data as typed XML objects that can better support interoperability among different applications. We plan to continue this work, incorporating XML/JAVA output facilities into other codes suitable for computational chemistry and biochemistry applications, such as molecular dynamics (e.g., AMBER [11]), Quantum Monte Carlo [12], and docking software. This method supersedes traditional methods we and many others in the computational chemistry community have applied: the use of 'cut-and-paste' or custom shell scripts to import and export data between applications. Additionally, we are working extensively with Chemical Markup Language, CML, integrating the data schemas into workflow systems, and providing bridges between the CML and the XML that we have specifically developed.

Component 3: The user interface, GEMSTONE, is a desktop application that provides a 'dynamic' user interface that updates automatically as additional services and/or applications become available. To accomplish this, an application shell, a local database/state manager, and various remote services and applications have been the focus of development. The interface provides the end-user with a portal into computational chemistry and biochemistry resources available on the worldwide grid. The GEMSTONE framework is intended to provide a built-in understanding of Computational Chemistry data types from molecules to proteins, and provides the ability to load, copy, edit, compute, and/or visualize these core data types. The set of core data types are those developed in the earlier efforts, components 1 and 2, as well as other data types that the Computational Chemistry community is developing, such as the Chemical Markup Language (CML) [13]. In addition, GEMSTONE provides access to a set of published services and workflows. Users can "load" GUI elements from the remote services as needed and execute service capabilities using the typed data previously loaded into the environment. Behind the scenes, as the user interacts with the GUI elements, GEMSTONE constructs a workflow that represents a generalization of user actions that can be saved and re-executed. Alternatively, the user can publish the recorded workflow in a central registry for others to continue discovery and usage. Because of this, the resulting structure becomes a window into new ways of thinking about our science and opens up opportunities for key developments in scientific research through the ability to connect resources and software.

Figure 1 shows the current GEMSTONE interface. On the left, the user can discover the set of services and workflows that have been created and published. The user can dynamically load GUI presentation elements from the remote services into the environment (center). The two side areas can be adjusted in size as needed. The design encompasses rich applications that execute directly on the desktop, thereby providing a high level of interactivity. This services-based infrastructure allows dynamic binding to available services and simple integration of new applications. Users typically interact with the GUI elements in ways that would result in the generation of a corresponding generalized workflow, representing the actual sequence of actions carried out. Workflow exploration enables users to explore data using a variety of available services and automatically generate corresponding workflows that can be saved or rerun. The publication of these user-designed workflows, which we plan to enable in the near future, will allow any workflows generated by one user to be shared by others.

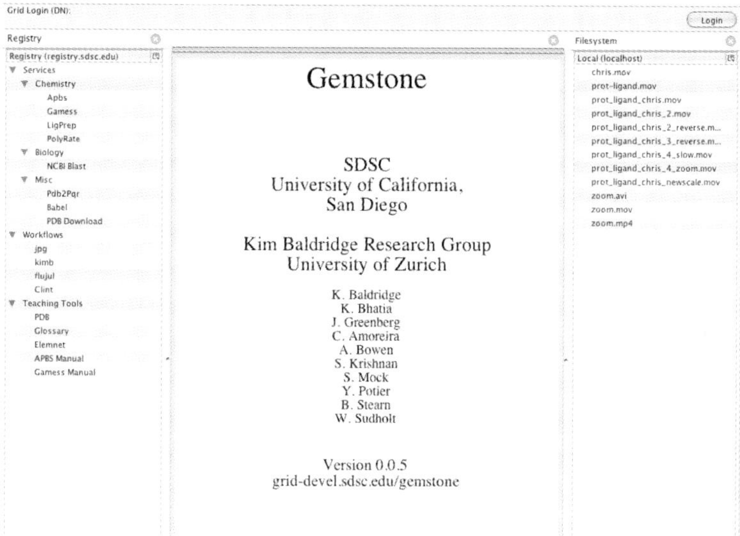

Figure 1.

The dynamic chemistry type system that we have created provides data validation and type checking for users as they interact with applications services. This is a particularly advantageous feature as it means that one does not have to spend an inordinate amount of time creating user-friendly interfaces that anticipates and corrects for any particular error that a user may generate through improper input/manipulation of the software. As such, the interface is automatically able to process and respond to user input and generate immediate feedback, and additionally, the issues of user-generated error do not have to be continually reinvestigated with each addition to the software.

The interface facilitates, through the InFORMNet component, integration and support for multiple computational resources, including high-end supercomputers, clusters or sets of clusters, and desktop grid systems. Access to databases, such as the protein database, as well as other peripheral resources can easily be made accessible through the interface.

3. LIFE SCIENCES DRIVEN USE CASES FOR GEMSTONE

It is envisioned that GEMSTONE will provide computational chemists and biochemists high flexibility within a dynamic environment to carry out various types of computational chemistry investigations. In particular, our wish was to create an exploratory environment where users can actually experiment with innovative arrangements of applications, services and data in a creative fashion – including ways in which are not anticipated by the creators of this environment. Furthermore, the environment should provide the capability to automatically record, save, and publish the corresponding workflow such that they, or any other user, can re-execute the saved workflow at a later date.

Together with grid resources, in particular we see such an environment as instrumental for several types of studies in Life Sciences, such as those that involve hybrid modeling efforts and/or high throughput studies. In this section we describe scientific investigations where we have begun to use GEMSTONE for such types of investigations.

Molecular and macromolecular studies: GEMSTONE provides a dynamic service explorer capability that provides a portlet-like environment for users to interact with remote services in such a way that supports innovative couplings between a wide range of different but related applications, including quantum chemical codes (e.g., GAMESS), classical codes (e.g., AMBER, APBS), molecular docking codes (e.g., AUTODOCK [14]), and hybrid modeling involving these same types of codes (e.g., QM/MM [15] studies). The different applications, while related, each represent a different algorithmic methodology and accuracy for understanding molecular structure and properties, and in many areas of research, these as well as other applications may be accessed by scientists in one of many combinations, not all of which can be anticipated by any one investigator.

As an example, consider molecular "docking," an important component of drug design. The goal in molecular docking is to estimate optimal three-dimensional configurations of complexes between proteins and ligands. This process usually involves extracting separate geometries for the protein and the ligand of interest from structural databases by scanning many proteins or ligands against each other.

The relative orientation between the protein and the ligand is varied until their optimal orientation is found, utilizing some specific structural or electronic criteria. To find the optimal steric fit corresponds to the scanning of both structures three translational and three rotational degrees of freedom as the ligand is 'adjusted' relative to the protein.

A major component of docking is to determine the underlying energy cost function to permit a judgment about which of the possible protein-ligand structures provides the best "fit". The cost function that defines the fit is usually assumed to consist of an electrostatic energy component, a non-electrostatic energy component, and an entropic component. The electrostatic binding energy between a protein and a ligand can be relatively accurately computed via dielectric continuum solvation methods based on the Poisson-Boltzmann equation. One can use the Adaptive Poisson-Boltzmann Solver (APBS) application for such calculations. The charge distribution on the ligand is determined using quantum chemical calculations, such as those provided by GAMESS, which provides highly accurate data for this structure. The two other components of the protein-ligand binding energy, the non-electrostatic and entropic components, are difficult to determine accurately by theoretical computations and thus often phenomenological equations are validated against real experimental results based on test sets where the correct protein-ligand structures are well known, for example from the Protein Data Bank (PDB) [16]. In addition, docking computations are likely to result in many suitable protein-ligand structures, and in this case, either a manual selection by the involved researcher or a rescoring by a more refined cost function is necessary.

In such an investigation, the end-user must be able to access multiple applications and databases. Each application typically has a complex set of execution parameters that must be varied through exploratory execution until the required values are determined. The GEMSTONE infrastructure directly supports such use by providing an environment where applications are automatically discovered and their GUIs loaded as needed. As well, the interface allows the user to smoothly operate between one application and another, with constant tracking of inputs and outputs, visualization capabilities at any step, and other data handling or archiving.

Consider the scenario in the figure to the right (Figure 2). The user initiates the GEMSTONE environment on a desktop or laptop computer. The data of interest is loaded directly into the GEMSTONE*: in this case, a protein molecule (prot587)

---

*If the data is large, the GEMSTONE environment simply maintains a reference to the data element stored at a remote location. All the capabilities of the GEMSTONE environment are maintained in this situation.

from the Protein Data Bank, and a ligand (ligand23) from a ligand database. These data items are strongly typed meaning that the data is well formed according to a specific XML schema describing the object. The GEMSTONE understands these schemas and will provide basic tools for viewing and editing the data once loaded into the environment. After the data is loaded, the user selects one or more set of services that are registered in the registry. All previously published workflows are described in the registry and are discoverable as are single web/grid services. In our example, the user selects the BABEL service to add hydrogens to the structures taken from the database, as these are typically missing. The user can drag and drop the files into the interface for this service, and the new files are generated and stored.

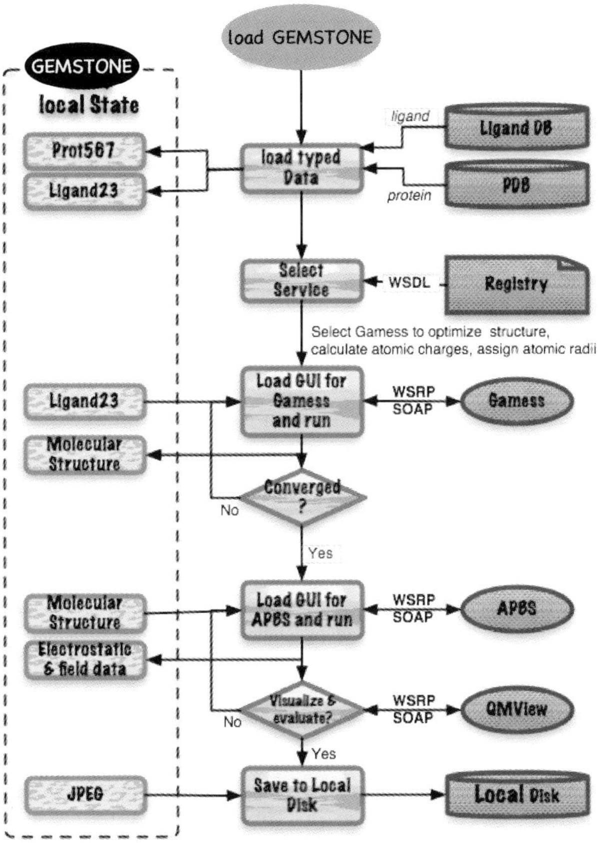

Figure 2.

At this point, it may be of interest to understand the relative binding energy for a variety of user-defined positions of the ligand inside the protein. In our example, it was of interest to change the position of the natural ligand in the pdb around two specific dihedral angles, and evaluate the sterics and electrostatics of that alignment in the protein. We have created a new service, called LigPrep, which allows the selection of the position manipulation followed by generation of the associated ligand structures and alignment back into the protein. In this case, we generate a series of ligands based upon rotations around the two specified dihedral angles, at 30-degree increments. We first obtain the sequence of ligand pdb files which appear in our directory, and subsequently we use our service to remove the natural ligand and replace the generated ligands one by one, to generate the new set of protein-ligand complexes to continue the investigation.

At this point, we now have a series of ligands with varying conformation, and the corresponding protein-ligand complexes. We now invoke the GAMESS service in order to calculate the charges of the ligand more accurately in preparation for classical electrostatic binding calculations on the protein-ligand complex. Once selected, the GEMSTONE dynamically loads the portlet-like GUI from the GAMESS service and the user drags and drops the ligand into it. One can then choose the appropriate keywords for the GAMESS run within the interface that appears in the workspace. The result of the calculation is an optimized molecular structure containing atomic coordinates, accurate atomic charges, and atomic radii. This data is also typed, automatically loaded into the GEMSTONE, and can be used in future calculations. If the results of the calculation are not sufficient, then the input to GAMESS can be modified and the calculations rerun.

After this stage, the user selects the APBS service for computation of the electrostatic field and binding energy of the protein-ligand complex, which will be invoked for each of the generated protein-ligand complexes. Again the GUI elements are dynamically loaded into GEMSTONE and the molecular structure and charges (the result of the GAMESS execution), together with the respective pdb files for ligand-protein complex and APBS specific input from the interface, is pulled together. The result of the computation that is submitted is a volume representing the electrostatic field and the electrostatic binding energy. The user can then invoke the QMView (or other visualization or analysis) service to view and subsequently convert to a JPEG format to saves the result.

Note that the work performed by the user could, in retrospect, be viewed as a simple data-oriented workflow (Figure 3a). In the near future, GEMSTONE will be capable of recording the users actions and building these workflows similar in nature to the “macro recording” capability found in many desktop applications.

These recorded workflows can be saved and re-executed later, or published and used by others.

Use of Pre-Defined Workflows for High-Throughput Processing: The first research example was relatively straight forward, involving a single user passing data between a few services. Another use of GEMSTONE is for more high throughput research setups. We assume first that a user has already explored the available applications and parameters to determine the set of applications needed and the parameters necessary for each application. Once known, an unrelated user may wish to execute that exact workflow on a very large number of molecules stored in a database, so called "high-throughput" studies. Note that here the workflow is quite complex in its implementation, but once it has been built and tested, it is far simpler for users to use the resulting workflow for new input data. In this case, the use of high-end grid resources is essential as the search space over optimal parameter values can be very high and can utilize significant computational cycles for execution.

As discussed previously, an important component of drug design is the calculation of the optimal binding between a protein and a potential drug (the ligand). For each protein-ligand complex, we wish to evaluate the orientation of the ligand relative to the protein and to calculate $\Delta E(protein\text{-}ligand) = E(protein\text{-}ligand) - E(ligand) - E(protein)$ where *ΔEprotein-ligand)* is the binding energy of the ligand minuts that of the protein complex, and *E(protein-ligand), E(ligand)* and *E(protein)* are the energies of the protein-ligand complex, the separate ligand and the protein, respectively. We can also define *ΔE(protein-ligand)* in terms of the following empirical formula: $\Delta E(protein\text{-}ligand) = \Delta E_{elst}\ (protein\text{-}ligand) + a\Delta E_{nonElst}(protein\text{-}ligand) + b$. Here, *a*, and *b* are adjustable parameters, and ΔE_{Elst}, $\Delta E_{nonElst}$, are the electrostatic energy and non-electrostatic binding energies respectively. For a particular complex, we can invoke workflow WF1, which would have been created by the user in the first use case above, employing GAMESS to minimize the ligand structure (if required) and associated charges, and subsequently passing the data to APBS to calculate the electrostatic energy component.

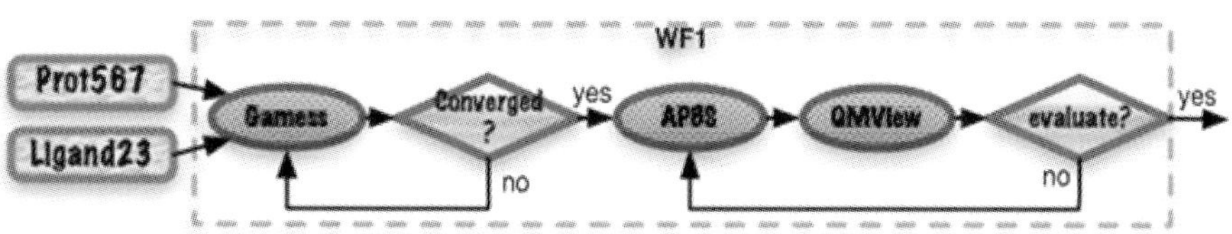

Figure 3a. Workflow WF1 (a) basic workflow created through the exploration interface in use case 1

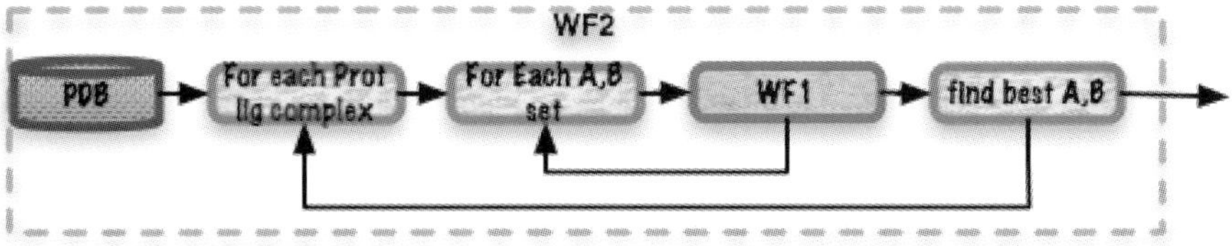

Figure 3b. Workflow WF2 (b) shows how workflows can be incorporated hierarchically

We use APBS via **WF1** to calculate the electrostatic energy of the isolated protein, and GAMESS/APBS to calculate the electrostatic energy of the isolated ligand (single calculations). Then, for given values of *a* and *b* we execute **WF2** (Figure 3b) to evaluate the optimal complexation of the protein-ligand by sweeping over their respective orientation space. Once one or more optimal orientations have been found, the workflow proceeds to the next (a,b) set. In the end, all (a,b) sets can be analyzed and compared.

This use case illustrates that workflows can be saved and published for re-execution by others and/or incorporated into other published workflows in a hierarchical arrangement. Here, **WF1** was saved, published and incorporated into **WF2**. As such, workflows can be effectively used for high-throughput processing.

This section outlined two science-based use cases that are driving our development and our software architecture. The system must support an explorative environment for accessing remote services, automatic generation of workflows based on user actions, publish and discovery capabilities for stored workflows, and integration with high-end grid resources for data management and computation.

4. ARCHITECTURE OF THE GEMSTONE USER INTERFACE APPLICATION

The overarching architecture for the GEMSTONE, shown below (Figure 4), is a layered architecture with resource services providing access to high-end data and compute resources; middleware services providing workflow execution capabilities; and the user environment services providing the user interface. The highlighted items include the GEMSTONE front-end, the workflow factory and engine, compute services and type system. The GEMSTONE user interface application is the user's *portal* into the grid: it is an end-user, cross platform java application running on the users desktop or laptop that provides access to the set of application services and workflows available to the user. Its core purpose is to:

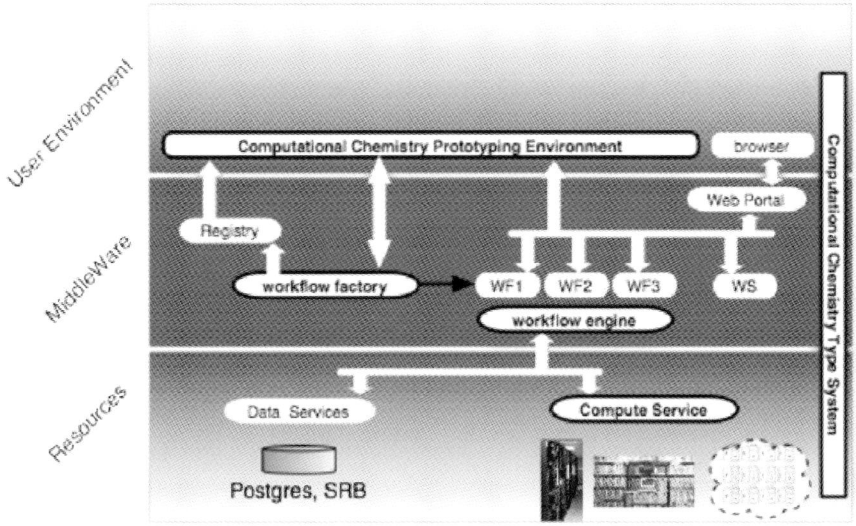

Figure 4.

a) Provide basic understanding of the type system and ability to view, edit and visualize typed data,

b) Support exploratory analysis by the scientist/end-user for manipulating domain-specific data, services and workflows,

c) Provide the capability to create, publish and manage workflows.

GEMSTONE provides a reusable presentation layer that can both take advantage of advanced graphical technologies as well as process interaction with graphical components of remote services. The architecture incorporates the Computational Chemistry type system (described further below) and provides native support for loading typed data into the GEMSTONE environment, for viewing and editing the data, and for basic visualization and management of the data. Additionally the loaded data can be used as input to the graphical components of the remote web services. Data type schemas, like those developed for GAMESS, can be incorporated in this fashion and provide the application the ability to use complex visualization and user interaction.

Based on our own experiences as well as our experiences with computational chemistry users, the interfaces that appear to be the most usable are those that provide an interface similar to common web-portals, providing interface components such as sliders for integer values, text boxes, and various buttons. Extensibility is needed to ensure the services and capabilities of the environment evolve to meet future needs of the scientists. Adding new capability to a portal requires extensive training and development work by the portal developer. In addition, the integration of new capabilities must be "open" and support the integration of third-party services and capabilities with limited developer support.

GEMSTONE provides a familiar "portlet"-style environment for users to interact with. Portlets are presentation or GUI fragments visualized by a portal container, typically a web server. In this traditional portal, the user connects via HTTP to a web server/portal container and retrieves a web page that contains the aggregated markup from various portlets. Frequently portlets are used to provide a human interface to a remote web service as in Figure 5a. Limited by HTML, the developer is unable to offer an interface that is visually compelling. Adding new capabilities to the portal requires additional portlets to be written. If a large number of web services are to be integrated, a significant development effort is required to construct all the corresponding portlets. As the remote service is updated the corresponding portlets require updating.

Recognizing this limitation, the web services community has been developing the WSRP specification (Web Services for Remote Portlets). Essentially, WSRP operates as a layer between portals and portlets so that each can be developed independently without changes to the other. Portals can dynamically find and add new portlets implemented by the web service developer (Figure 5b). As web services are updated, the corresponding portlets are automatically reloaded. Additionally, because GEMSTONE operates as the end user's "browser" as well as a portal, the HTML bottleneck is removed; portlets can take full advantage of the GEMSTONE presentation layer when defining an interface.

Another advantage of the portlet-based user interaction is that the actions of a user can easily be captured and recorded for later playback. As users load data from data storage into the GEMSTONE environment and invoke various applications or services, new typed data is generated and automatically added into the environment. The user's actions on a set of data form a directed acyclic graph (DAG) that defines a partial order on the sequence of applications/services executed. The GEMSTONE environment will record the actions and maintain the DAG. The user can view the generated DAG, reload and edit data and replay previous actions as needed. The generated workflow can also be saved and published by the user so that other users of the system can discover and execute the same workflow, or embed the published workflow into new workflows.

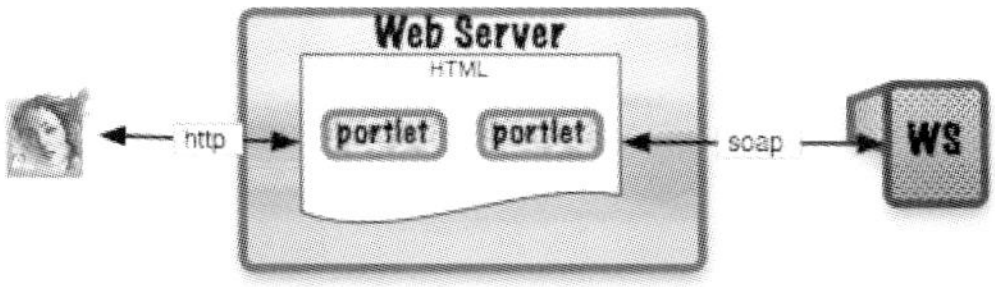

(a)

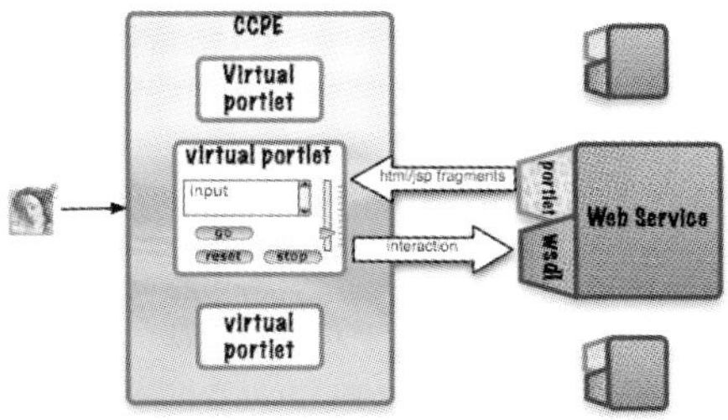

(b)

Figure 5. Typical portlet-based web portals (a) and WSRP-based portlets (b)

5. COMPUTATIONAL CHEMISTRY TYPE SYSTEM

The most popular way to process and analyze data obtained from computational software is to directly manipulate the text of the output using shell scripts or even by cutting and pasting from output files. However, there are inherent problems in this approach. Human error resulting from the construction of input files by hand (particularly complex ones), and manipulation of output files is always a factor, and for example, a change in the output format of the software program may result in a failure of the script to process the output correctly. Instead, it is more efficient to put structured data output facilities directly into the computational program. We have adapted the methods of data representation and transport used in the grid and web services that make up the Scientific Workflow to process data obtained from computational chemistry codes. We have designed an XML schema based on the output of the GAMESS quantum chemistry code [3, 17-18], and have used castor [19] to generate JAVA code that allows for the processing of the data in the format specified by the schema. Representation of data via XML documents allows for identification of the data via the tag names (as opposed to static tag names defined by HTML for example), the establishment of a hierarchy of elements that mimic the natural data structure, and serialization of data into java objects.

Similar adoption of structured input and output can be incorporated into other computational codes, potentially opening up opportunities for hybrid integration. We plan to continue this work, incorporating XML/JAVA output facilities into code such as APBS, as well as auxiliary codes that enable processing of data between vintage software codes. Though this work can be more laborious initially than parsing output files, it is well worth the effort as the data is put into a form that may be used for all subsequent use of the software, database storage, querying, and efficient transport over the grid in the form of JAVA objects.

The current schema may be viewed at http://www.sdsc.edu/~jpg/nmi/gamess.xsd. The schema contains the essential data that is output from a quantum chemistry code, such as atomic coordinates, molecular orbital data, and first and second derivatives of the energy with respect to the coordinates. Work continues to develop this schema and the codes that support it. All GAMESS input options are included in the schema and XML/JAVA input routines put into GAMESS so that XML files/JAVA objects may be read as input. This facilitates a wider range of workflows. Additional schemas are also being designed to provide compound identification and job information.

The project described above is a first step in obtaining such standard data types. Beyond the adoption of data processing and archiving procedures for GAMESS and the associated programs or even other QC programs, is the idea of devising uniform data types that cut across particular disciplines. That is the ability to pass data from an *ab initio* QC program to, for example, a molecular mechanics program. Uniform data types would greatly facilitate both programming connections between different codes, storing the data in databases and promoting future collaborations with researchers in related, but possibly different disciplines. Work has been done in this area by others, most notably in the development of CML (Chemical Markup Language) [13], its schema [20] and the associated CMLDOM [26]. CML provides a comprehensive format for describing chemical data in terms of XML documents, and as such provides bridge technology for interoperability between the various QC software CML is general enough so that elements may be readily extracted and schemas built for specific projects. The CML approach has a modified way to parse and access data from JAVA objects. The central part of our data structure is the GAMESS schema, designed specifically for QC data and GAMESS input options, from which we generate data bound java objects: that is the java classes and methods refer directly to the data that they reference. The Document Object Model (DOM) represents the data as a tree structure and uses abstract classes and methods that refer to the element, its attributes, and children rather then directly to the data. Additional JAVA classes may be built around these classes to reference the data directly. In fact, the two approaches will be highly complimentary and it is expected that CML will provide such a bridge in the case of other QM software in the same manner. A schema derived from CML may be represented in terms of data bound objects by Castor, or a DOM object may be created from a GAMESS derived XML document. The data, as represented in both approaches may also, readily be transferred between formats.

6. FUTURE DEVELOPMENTS OF GEMSTONE

The workflow component of the GEMSTONE is a direct evolution from the workflow component built as part of the prior work, called Informnet, and is the core component that ties together the frontend with the computational and data resources. Informnet already provides a mechanism for creating workflow instances through the workflow factory (Figure 4), and the means of executing workflows through the workflow engine. Informnet provides GSI [21] authentication and security and integrates with basic computational and data resources using Globus and GridFTP [22].

A mechanism for publishing and discovering published workflows is planned for our future versions of GEMSTONE. Once a workflow has been specified and tested, a researcher may wish to share this workflow with others as a service. GEMSTONE will facilitate the exploration and publishing of a pre-defined workflow into Informnet. The Informnet system will then instantiate a copy of these pre-defined workflow services for use by others. Another service will maintain a list of the workflows, their XML specifications, and a description of the functionality provided by a published workflow. The discovery service will be revealed in GEMSTONE within an interface that can be queried for a list of services and the accompanying information associated with the services.

Enhanced fault tolerance and fault reporting will also be added. Use of priority-based notifications sent to the clients and recovery of failed executions provide the user the ability to respond to a fault by changing the workflow description and continuing. This will require the development of pause and rewind capabilities in the workflow engine service API exposed to the client. If a client receives a fault notification, the client may choose to pause or rewind the workflow and reconfigure any number of the tasks, then send the workflow engine the reconfigured workflow and the restart command.

More tasks types will be incorporated to support the Chemistry Environment. Informnet currently supports the execution of tasks as executables local to the computational resource running the Informnet service, as well as on any computer on the grid running Globus. The task types can be expanded to include other web services and grid services to enable the use of external services that the scientist has discovered. The workbench will maintain a list of services that the user can select from to build their instance of a workflow.

In addition to more task types, more file protocols will be added to the architecture for automatic staging and archival of input and output from the task execution. Informnet currently supports local filesystems and GridFTP. The addition of FTP, HTTP, SCP, and SRB [23] would greatly add to the flexibility and reach of the Informnet system.

The workflow description language supports associating more than one computational resource as the endpoint to run a specific task. This flexibility was created to enable brokering of tasks within the Informnet architecture. A broker could query information sources like the MDS[24], INCA[25] or any other information service and perform late binding of the task to a computational resource based on the information gathered. The brokering software still needs to be developed, as do examples of brokering policies to be plugged into the architecture.

7. CONCLUSIONS

The described research involves development and deployment of an integrated and practical grid architecture for computational chemistry and biochemistry research. Our preliminary work has resulted in a prototype dynamic workflow driven system, coupling domain specific workflow for interoperations with computational chemistry and biochemistry software, with general execution workflow for handling all aspects of job submission, execution and monitoring automatically. The ongoing efforts drive towards significantly improving on the coupling and interoperability with added resources, and make a natural evolution into a dynamically configurable user interface. Such a coupled capability for interactively steered and adaptive resource connection enables real-time construction of creative workflow strategies. The resulting Grid-Enabled Molecular Science Through Online Networked Environments (GEMSTONE) infrastructure encompasses interactive application space, dynamic binding to services and new applications, workflow exploration and publication, and accommodation to a variety of compute options. Recent advances in grid infrastructure and associated protocols enable the pioneering of such capability. The overall objective of this work is the development of an innovative grid architecture integrating real-time applications, data and intervention, and provisions for post-computing analysis and modeling of data and image repositories. The resulting information technology (IT) research is expected to have broad impact on the computational (bio)chemistry/biophysics research communities and knowledge base in terms of methods usage and development for practical problems, as the technology is readily adaptable.

The intellectual merit of this work lies in the a) IT research questions that will be answered for the applications area, b) scientific research questions that will be enabled by development of the dynamic work environment, c) ability to revolutionize the way researchers think and perform research in such an innovative working environment, and c) capacity of GEMSTONE to influence the creation of new strategies and algorithmic possibilities for scientific research. The IT contributions will include the development of domain-specific abstractions to access grid resources, and the development of technologies that merge the usability of web portals with the flexibility of workflow systems. Web-based portals are familiar and ubiquitous interfaces for novice end-users, however, they lack the flexibility for expert-users and do not support the exploratory process enabled by workflow systems, which provide a high degree of flexibility and enable innovative arrangements of applications, services and data. The combination of the strengths of each technology is expected to be quite powerful. The proposed infrastructure uses a unique mix of client-side and server-side systems leveraging web-services infrastructure, a key engineering challenge.

The broader impact of this work revolves around the ability to deploy such capability to both computational modeling and design experts, as well as non-experts who wish to exploit aspects of the tools without being masters in the field. The proposed infrastructure will promote scientific discovery within and across virtual organizations, benefiting many aspects of science and development. Currently, infrastructure at the level we propose does not exist. We will evaluate, develop and apply technology to deliver dynamic computational science resources as grid services. Since the mechanisms and protocols resulting from this effort will be generally applicable to many disciplines, software will be openly distributed for education, research and non-profit purposes. The project moreover offers substantial opportunities for training the next generation of computer and applications scientists (in particular the students who will be funded off of this grant), and should be extensively beneficial for classroom instruction.

ACKNOWLEDGMENTS

We gratefully acknowledge in particular the National Science Foundation National Middleware Initiative, ANI-0223043, and the University of Zürich, for funding of this work. We acknowledge scientific motivations from the NIH-NBCR-RR08605, and grid community development motivations from PRAGMA. We thank PRAGMA, NIH-NBCR, UZ, and SDSC, who may have provided computing resources for any grid testing on applications, and their system administrators, in particular the SDSC ROCKs group.

REFERENCES

1. Papadopoulos, P., K. Baldridge, and J. Greenberg, Integrating Computational Science and Grid Workflow Management Systems To Create a General Scientific Web Service Environment, 2002, NSF award #ANI-0223043.
2. Informnet, http://chemport.sdsc.edu/informnet/.
3. Baldridge, K.K., J.P. Greenberg, S. Elbert, S. Mock, and P. Papadopoulos. QMView and GAMESS: Integration into the World Wide Computational Grid. in SuperComputing 20022002. Baltimore, MD.: IEEE Computer Society Press.
4. Greenberg, J., S. Mock, M.J. Katz, G. Bruno, F.D. Sacerdoti, P. Papadopoulos, and K. Baldridge. Incorporation of Middleware and Grid Technologies to Enhance Usability in Computational Chemistry Applications. in ICCS 042004. Krakow, Poland: Springer - Verlag.
5. Schmidt, M., K.K. Baldridge, J.A. Boatz, S. Elbert, M. Gordon, J.H. Jenson, S. Koeski, N. Matsunaga, K.A. Nguyen, S.J. Su, T.L. Windus, M. Dupuis, and J.A. Montgomery, The General Atomic and Molecular Electronic Structure System. Journal of Computational Chemistry, 1993. **14**: p. 1347-1363.
6. The Physiology of the Grid: An Open Grid Services Architecture for Distributed Systems Integration, 2002, http://www.globus.org/research/papers/ogsa.pdf
7. Baldridge, K.K. and M.W. Schmidt, 3D-PLTORB, 1997 San Diego.
8. Baldridge, K.K. and J.P. Greenberg, QMView: A Computational 3D Visualization Tool at the Interface Between Molecules and Man. Journal of Molecular Graphics, 1995. **13**: p. 63-66.
9. Baldridge, K.K. and J.P. Greenberg, QMView as a SupramolecularVisualization Tool, in Supramolecular Chemistry, J. Siegel, Editor. 1995, Kluwer Academic Publishers. p. 169-177.
10. Baker, N., M. Holst, and F. Wang, Adaptive multilevel finite element solution of the Poisson-Boltzmann equation II. Refinement at solvent-accessible surfaces in biomolecular systems. Journal of Computational Chemistry, 2000. **21**(15): p. 1343 - 1352.
11. D.A. Pearlman, D.A. Case, J.W. Caldwell, W.R. Ross, T.E. Cheatham, III, S. DeBolt, D. Ferguson, G. Seibel and P. Kollman. AMBER, a computer program for applying molecular mechanics, normal mode analysis, molecular dynamics and free energy calculations to elucidate the structures and energies of molecules. Comp. Phys. Commun. **91,** 1-41 (1995)

12. Kim, J., Ceperley, D. M., Computational Materials Group, University of Illinois, Champaign.
13. Murray-Rust, P., H.S. Rzepa, and M. Wright, Development of chemical markup language (CML) as a system for handling complex chemical content. New Journal of Chemistry, 2001(4): p. 618-634.
14. Morris, G.M., D.S. Goodsell, R.S. Halliday, R. Huey, W.E. Hart, R.K. Belew, and A.J. Olson, Automated Docking Using a Lamarckian Genetic Algorithm and Empirical Binding Free Energy Function. J. Computational Chemistry, 1998. **19**: p. 1639-1662.
15. Gao, J., Review on QM/MM. Reviews in Computational Chemistry, 1996. **7**: p. 119-185.
16. The Protein Data Bank, http://www.pdb.org.
17. Welch, V., F. Siebenlist, I. Foster, J. Bresnahan, K. Czajkowski, J. Gawor, C. Kesselman, S. Meder, L. Pearlman, and S. Tuecke. Security for Grid Services. in Twelfth International Symposium on High Performance and Distributed Computing2003. Seattle, WA.
18. Allcock, B., J. Bester, J. Bresnahan, A.L. Chervenak, I. Foster, C. Kesselman, S. Meder, V. Nefedova, D. Quesnal, and S. Tuecke, Data Management and Transfer in High Performance Computational Grid Environments. Parallel Computing Journal, 2002. **28**(5): p. 749-771.
19. Rajasekar, A.K. and M. Wan. SRB and SRBRack- Components if a Virtual Data Grid Architecture. in Advanced Simulation Technologies Conference. April 15-17 2002. San Diego CA.
20. Zhang, X., J. Freschl, and J. Schopf. A Performance Study of Monitoring and Information Services for Distributed Systems. in Twelfth International Symposium on High Performance and Distributed Computing2003. Seattle, WA.
21. San Diego Supercomputer Center, The Inca Test Harness and Reporting Framework, 2004,
22. Baldridge, K. and J. Greenberg. Management of Web and Associated Grid Technologies for Quantum Chemistry Computation. in ICCS 032003. Melbourne Australia.
23. Greenberg, J.P. GAMESS/QMVIEW. in Computational Representation of Bio-Molecules , 20032003. UCSD: Michele Sanner.
24. Castor, 2002, The Exolab Group.
25. Murray-Rust, P. and H.S. Rzepa, Chemical Markup, XML, and the World Wide Web. 4. CML Schema. Journal of Chemical Information and Computer Science, 2003. **43**: p. 757-772.

APPLICATION-LEVEL QOS SUPPORT FOR A MEDICAL GRID INFRASTRUCTURE*

SIEGFRIED BENKNER, GERHARD ENGELBRECHT, IVONA BRANDIC, RAINER SCHMIDT

Institute of Scientific Computing, University of Vienna,
Nordbergstraße 15, Vienna, Austria
Email: {sigi, gerry, brandic,rainer}@univie.ac.at

STUART E. MIDDLETON

IT Innovation Centre, University of Southampton, Southampton, UK
Email: sem@it-innovation.soton.ac.uk

Quality of Service is a crucial issue in the context of providing medical applications on the Grid. The GEMSS Project, which develops a Grid infrastructure for medical simulation services, addresses this issue by providing application-level Quality-of-Service support in the form of explicit timeliness guarantees for time-consuming simulation services. The GEMSS infrastructure and middleware are based on standard Web services technology ensuring future extensibility and interoperability. Within GEMSS, parallel applications installed on clusters or other HPC hardware may be exposed as QoS-aware Grid services, which are capable of dynamically negotiating with clients QoS constraints for response time and price. QoS constraints agreed upon between client and service providers are expressed based on the Web Service Level Agreement specification. In this paper, we present an overview of the GEMSS infrastructure, outline the provision of parallel MPI codes running on clusters as Grid services, and describe the GEMSS QoS infrastructure in more detail.

1. INTRODUCTION

The GEMSS Project[14] is developing a secure, service-oriented Grid infrastructure for the provision of advanced medical simulation applications as Grid services. The medical prototype applications considered in GEMSS include maxillo-facial surgery simulation[19], neuro-surgery support[28], radio-surgery planning[13], inhaled drug-

*The work was supported by the European Union's GEMSS Project under contract IST 2001-37153 and by the Austrian Science Fund as part of the AURORA project under contract SFB F011-02.

delivery simulation[17], cardio-vascular simulation[18] and advanced image reconstruction[2]. At the core of these bio-medical simulation applications are computationally demanding methods such as parallel Finite Element Modeling, parallel Computational Fluid Dynamics and parallel Monte Carlo simulation, which are realized as remote Grid services running on clusters or other parallel computing platforms. To ensure the use of these services in a clinical environment, predictability of response times of remote simulation services is of utmost importance. To address this issue, we have developed a flexible Quality of Service (QoS) infrastructure for providing explicit response time guarantees for simulation services which are executed remotely on some GEMSS Grid host. Response time guarantees are usually negotiated dynamically between a client and potential service providers on a case-by-case basis. The QoS infrastructure is generic in the sense that arbitrary QoS parameters may be supported. Since GEMSS also addresses the realization of Grid business models, services may be configured to support dynamic price negotiation as well. QoS guarantees agreed between a service consumer and a service provider are expressed in form of an XML document following the Web Service Level Agreement[34] (WSLA) specification. Besides explicitly negotiable QoS guarantees, the GEMSS infrastructure provides implicit QoS by realizing highest security levels and providing support for error recovery.

The GEMSS Grid infrastructure and middleware have been built on top of standard Web services technologies,[30,33,36] ensuring future extensibility and interoperability. Furthermore, GEMSS addresses privacy, security and other legal concerns by examining and incorporating into its Grid services the latest laws and EU regulations related to providing medical services over the Internet[24].

In this paper we present an overview of the GEMSS infrastructure, outline the provision of parallel simulation codes running on clusters as Grid services, and describe the QoS support infrastructure in more detail. The remainder of this paper is organized as follows: Section 2 presents an overview of the GEMSS Grid infrastructure and discusses the provision of applications as Grid services. Section 3 describes the GEMSS QoS infrastructure and the basic strategy for QoS negotiation. Section 4 presents a case study of a medical image reconstruction service. Finally, a discussion of related work and conclusions are presented in Sections 5 and 6, respectively.

2. GEMSS GRID ARCHITECTURE AND INFRASTRUCTURE

The GEMSS infrastructure is based on a service-oriented architecture comprising multiple clients and services, one or more service registries, and a certificate authority. Service registries maintain a list of service providers and the services they support. The certificate authority provides the basis for an operational PKI infrastructure based on X.509 certificates for establishing an identity for clients and service providers as well as for realizing transport and message layer security.

Grid Clients are usually Internet-enabled PCs or workstations with GEMSS client software installed that permit communication with a service provider through the GEMSS middleware. The client side applications handle the creation of service input data and visualization of service output data.

GEMSS services encapsulate native HPC applications (usually parallel simulation kernels written in Fortran or C and MPI) and provide support for quality of service negotiation, data staging, job execution, job monitoring, and error recovery and are usually accessed subject to a chosen business model. GEMSS services are defined via WSDL and securely accessed using SOAP messages. For large file transfers, SOAP attachments are utilized. Since GEMSS supports a client driven approach for accessing services, it is not required that holes be tunneled through site firewalls. End-to-end security is realized on top of transport layer security (HTTPS, SSL) and message layer security utilizing WS-Security standards[36].

GEMSS supports a three-step process to job execution. First there is an initial business step, where accounts are opened and payment details fixed. Next there is a quality of service negotiation step, where a job's quality of service and price, if not subject to a fixed price model, is negotiated and agreed. Finally, once a QoS contract is in place, the job itself can be submitted and executed.

2.1. Service Provision Infrastructure

Figure 1 (right) depicts the main architectural components of the GEMSS Grid service provision infrastructure. Medical simulation applications are exposed as Web Services and hosted using a Web server and a service container (currently Apache and Tomcat Axis). The quality of service management component handles reservation with the compute resource manager (job scheduler) and provides input to the quality of service negotiation process. The error recovery component handles check pointing and re-starting of applications if required. The logger manages a database for logging auditable information and a low level system log for event logging. The service state repository component manages a state database that contains information about any client-service interaction allowing it to be resumed at a later time if the user logs off. The provision of applications as Grid services is based on the concept of generic application services and described in more detail in Section 2.3.

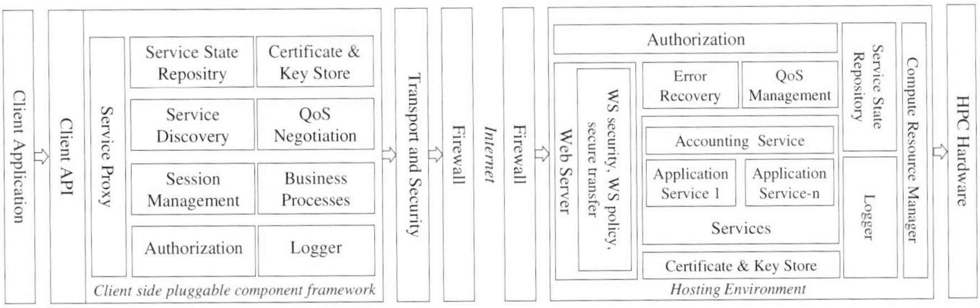

Figure 1. GEMSS client (left) and service provider infrastructure (right).

2.2. Client Infrastructure

The main architectural components of the GEMSS Grid client infrastructure are shown in Figure 1 (left). The client-side application code usually relies on the GEMSS client application programming interface (API) which hides most of the complexity of dealing with remote services from the application developer by providing appropriate service proxies. Service proxies are in charge of discovering services and negotiate with Grid services to run jobs on its behalf. The session management component manages client sessions, and a security context is maintained allowing authentication of the current user and providing the access criteria for the certificate and key stores. A service discovery component is provided for looking up suitable services in a service registry. The client typically runs a business workflow to open negotiations with a set of service providers for a particular application. The quality of service negotiation is then run to request bids from all interested service providers who can run the clients jobs subject to QoS criteria required by the client; this results in a contract being agreed with a single service provider. The client then uploads the job input data to the service provider and starts the server side application by calling appropriate methods of the service. The client infrastructure is centered on a pluggable client side component framework which provides support for dynamic configuration and replacement of client-side components.

2.3. Generic Application Services

A major objective of the GEMSS project was the development of a generic Grid service provision framework that simplifies the task of transforming existing applications into Grid services without having to deal with the details of Web services and Grid technologies.

The transformation of medical simulation applications into Grid services is based on the concept of generic application services[4]. A generic application service is a configurable software component which exposes a native application as a service to be accessed on demand by multiple remote clients over the Internet. A generic application service provides common methods for data staging, remote job management, error recovery, and QoS support, which are to be supported by all GEMSS services. In order to customize the behavior of these methods for a specific GEMSS application, an XML application descriptor has to be provided. The application descriptor specifies the input/output files, the script for initiating job execution, and a set of performance-relevant application parameters required for QoS support. A generic application service is realized as a Java component, which is transformed automatically into a Web service with corresponding WSDL descriptions and customized for a specific GEMSS application using the XML application descriptor.

2.4. Application Requirements

In order to provide a native application as a GEMSS Grid service, the application has to be installed on some Grid host and a job-script to start the application as well as an XML application descriptor have to be provided. For QoS support, usually a performance model to estimate the response time and a pricing model for determining the price of a job execution has to be provided (see Section 3 for more details). Usually no code changes are required, provided the application can already be executed in batch mode and that files in I/O operations are not accessed with absolute path names.

2.5. Service Deployment

A generic application service encapsulating a GEMSS simulation service has to be deployed within the GEMSS hosting environment which currently relies on Apache Tomcat/Axis. As a result of deployment, the native application is embedded within a generic application service and accessible over the Internet. In order to provide support for automatic service deployment, a corresponding GEMSS deployment tool has been developed. The deployment tool enables the user to enter the information required in an application descriptor via a GUI and to control the deployment process. Internally, the deployment tool creates the XML application descriptor, generates an appropriately customized Web service which encapsulates the application, publishes the corresponding WSDL document in a registry service, and finally deploys the service within the GEMSS hosting environment.

3. QOS SUPPORT

Using the GEMSS infrastructure, medical simulation applications available on clusters or other parallel hardware may be exposed as QoS-aware services which are capable of negotiating with clients QoS guarantees on execution time, price and others. The GEMSS QoS negotiation mechanisms enable a client to negotiate with one or more service providers the required end-time at which the results of a time-critical simulation job have to be ready. Service providers utilize machine-specific application performance models in order to estimate the required execution time for a specific job based on input meta-data supplied by the client during QoS negotiation. In order to ensure the availability of appropriate computing resources for a service request, the GEMSS service provision environment relies on a scheduling system that provides support for advance reservation.

3.1. QoS Infrastructure

Figure 2 presents the main parts of the GEMSS QoS infrastructure separated into client-side and service-side parts.

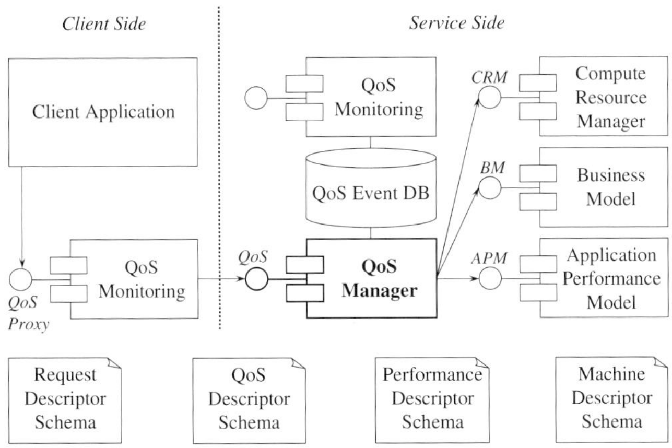

Figure 2. GEMSS QoS infrastructure.

On the client-side a QoS component is provided, which offers the *QoS Proxy* interface to be utilized by GEMSS client applications for QoS negotiation. The QoS proxy interface provides methods for requesting, confirming and for canceling a QoS contract, which are used during QoS negotiation.

The service-side QoS management module is centered on the QoS manager, which interacts with the compute resource manager (CRM), the application performance model (APM) and the business model (BM). The QoS infrastructure relies on four different XML schemata for the specification of QoS descriptors, request descriptors, performance descriptors and machine descriptors. The QoS event database is utilized by the QoS manager and the QoS monitoring service to store and query specific QoS events.

3.2. QoS Management Module

The QoS manager is the central server-side module of the QoS support infrastructure and provides the interface *QoS* with basic QoS negotiation operations. The QoS manager receives a QoS request from a client, checks whether the client's QoS constraints can be met, and generates a corresponding QoS offer which is returned to the client.

The QoS manager utilizes the application performance model (APM) to estimate the performance requirements (runtime, memory and disc requirements). The performance model takes as input a request descriptor and a machine descriptor and returns a performance descriptor. A request descriptor contains meta-data about a specific service request (e.g. mesh size, image resolution, etc.) supplied by the client during QoS negotiation. A machine descriptor contains machine specific information, usually specifying the number of processors, or a range of feasible processor numbers that should be used for executing an application. A performance descriptor comprises information on job capacity estimations including runtime, disc requirements and memory requirements. Since in general, it will not be possible to build an analytical model which allows for precise predictions of memory and computing time requirements for all applications, GEMSS does not prescribe the nature of a performance model. For applications where an analytical performance model is not feasible, for example, a data base relating typical problem parameters to resource needs like main memory, disk space and runtime, which will initially be populated using data from test cases, could be used.

The compute resource manager (CRM) module realizes a high level interface to an underlying scheduling system which has to provide support for advance reservation (e.g. NEC's COSY scheduler[9] or the Maui scheduler[22]). The CRM module provides methods for requesting and for confirming temporary reservations, for canceling reservations, for job submission and for inquiring information about a submitted job.

The service provider's business model defines a generic mechanism to calculate a price based on estimated resource allocation. A concrete business model implementation has to be provided by the service provider.

3.3. Basic QoS Negotiation

The basic QoS negotiation in GEMSS is based on a request-offer model where the client requests offers from service providers. If the client agrees to an offer, it is confirmed by the client and signed by both parties resulting in a QoS contract. The details of the basic QoS negotiation process are explained in the following.

(1) In an initial task the client may access a GEMSS registry service to obtain a list of potential service providers to be contacted during QoS negotiation. (2) The client generates a request descriptor and a QoS request which are sent to the potential service providers. The request descriptor contains meta data about the client request (e.g. mesh size, resolution, etc.) and the QoS request specifies the required QoS properties (maximal cost, earliest start of job execution, latest finish time). (3) On the server side the QoS manager attempts to generate a QoS offer by executing the performance model and business model and by checking with the compute resource manager if the required resources can be made available. (4) Based on the estimations and available resources, the QoS manager performs a temporary resource reservation with a short expiration time and returns a QoS offer to the client. (5) On the client side, the QoS offers from different service providers are received and analyzed. (6) The client confirms the best offer, or, if it is not satisfied with the offered QoS constraints, may set up a new QoS request with different constraints and continues with step 1. (7) On the server-side, the QoS manager confirms the temporary resource reservation made for the offer, signs the QoS contract and returns it to the client. (8) After the basic QoS negotiation the regular job-handling workflow is initiated. This usually comprises uploading of input data, starting of the native application, and downloading of results.

Within the GEMSS project also more sophisticated negotiation strategies based on auction models have been realized, a description of which is beyond the scope of this paper.

3.4. QoS Descriptors — WSLAs

A QoS descriptor is an XML-based document representing a (potential) agreement on a single service usage between a service consumer (client) and a service provider, following the Web Service Level Agreement (WSLA) specification[34]. The WSLA specification is a de-facto standard specified by IBM in 2003 for agreements on Web service usage among a set of involved parties. The GEMSS QoS infrastructure utilizes a subset of the features defined in the WSLA specification.

Depending on the state of a QoS negotiation, a QoS descriptor either is a QoS request, a QoS offer, or a QoS contract. QoS descriptors consist of three main parts: parties, service definition, and obligations. The parties section comprises information about the signatory parties involved, which is usually extracted from the GEMSS certificates of users and service providers, respectively.

The service definition section contains the actual subject matter of the agreement by defining all operations subject to the agreement and a set of Service Level Agreement (SLA) parameters. In the context of GEMSS, SLA parameters usually include the begin time of the job execution, the end time of the job execution, and the price of the job execution. Furthermore the service definition section specifies the overall contract duration and a metric for each parameter.

The obligations section contains a list of objectives. Each objective is linked to an obliged party, has an according validity and defines an expression that is associated with a defined SLA parameter. For example, the SLA parameter price has to be equal to 5 EUR or the end time of the job execution must not exceed 19 May 2005, 11:00 CET.

4. CASE STUDY — MEDICAL IMAGE RECONSTRUCTION

This section discusses experimental results of a performance model for a medical image reconstruction service for single photon emission computer tomography (SPECT). The parallel reconstruction kernel applies a compute-intensive fully 3D ML-EM reconstruction algorithm[2], which is implemented in C/MPI. The reconstruction kernel has been made available as a GEMSS service on two different 64-processor PC clusters. The parallel reconstruction kernel exhibits a good scaling behavior up to the full size of the clusters in our test environment. The client side application has been written in Java and provides an ImageJ-based GUI for the acquisition of 2D images, for the specification of reconstruction parameters (e.g. region of interest, required accuracy) and for the display of reconstructed 3D images. In the experimental setup, the size of the input images ranges from 1 to 4 megabytes and the output data size was about 2 megabytes. The average runtime of our tests was between 4 and 20 minutes and the results have shown that the overhead introduced by the QoS negotiation is below 10 percent.

Table 1 shows the information to be supplied by the client application in a request descriptor. The request descriptor contains meta-information about the actual input data and reconstruction parameters which are automatically extracted at the client-side from the input data. During QoS negotiation, the request descriptor is sent to the service and fed into the application performance model.

Table 1. Information contained in request descriptor (left) and performance descriptor (right).

Nr. of iterations	10	Radius of model	350
Nr. of slices	32	Matrix threshold	10e-6
Resolution	128,128	Model slope	0.0036
Nr. of projections	120	Voxel size	4.42

Nr. of processors	16
Disc space (mb)	627
Runtime (secs)	366
RAM (mb)	164

The performance model delivers a performance descriptor (Table 1) containing estimates for the execution time, memory requirements, and disc space. Moreover, the number of processors used for executing the reconstruction kernel is determined by the QoS manager such that the client's QoS constraints can be met.

Table 2. Image reconstruction service: Average measured runtime vs. estimated runtime.

Input	# Runs	Av. Time	Est. Time	Accuracy	Max. Dev.	Std. Dev.
Dataset 1	10	259	278	0.93	22	9.74
Dataset 2	12	437	512	0.92	40	19.80
Dataset 3	58	627	665	0.94	40	18.97
Dataset 4a	66	645	662	0.97	20	8.63
Dataset 4b	87	677	691	0.98	26	9.05
Dataset 4c	3	1132	1159	0.98	28	14.62

Table 2 presents a comparison of estimated runtime vs. real runtime of four different input datasets. The input datasets 1 - 4 vary in image resolution, computational accuracy (number of iterations) and model accuracy. The runtime of the equal datasets 4a - 4c is different because, depending on the client's QoS requirements, different numbers of CPUs have been chosen by the QoS manager. The average runtime for a single job (3rd column) was derived from a number of successive runs (2nd column). The 4th column shows the estimated runtime computed by the application performance model. Furthermore, the average accuracy (5th column), the maximal deviation of all runs (6th column) and the standard deviation (7th column) are shown.

The experimental results show, that even for a highly sophisticated medical image reconstruction algorithm an adequate performance model can be developed to fulfill the requirements of the GEMSS QoS support infrastructure.

5. RELATED WORK

There are a number of other Grid projects that deal with bio-medical applications, including the EU BioGrid Project[5], the OpenMolGRID Project[25], EU MammoGrid Project[20], and the UK e-Science myGrid Project[21]. While most of these projects focus on data management aspects, the GEMSS project focuses on the computational aspect of the Grid, with the aim to provide hardware resources and HPC service across wide area networks in order to overcome time or space limitations of single HPC systems. Other projects in the bio-medical field which also focus more on the computational aspect of the Grid include the Swiss BiOpera Project[6], the Japanese BioGrid Project[16], and the Singapore BioMed Grid[7]. However, none of these Projects addresses the issues of application-level QoS support.

The work presented in[23] deals with a QoS based Web services architecture. The system consists of QoS-aware components which can be invoked using a QoS negotiation protocol. As opposed to our work, this system does not deal with the Grid-provision of HPC applications. Several projects have proposed economy-based Grid systems[8,15]. Buyya proposed a Grid Architecture for Computational Economy (GRACE) providing a generic way to map an economic model into distributed system architecture and the Grid resources broker (Nimrod-G) supporting deadline and budget based scheduling of Grid resources.

The GRIA[26] QoS infrastructure utilizes a performance estimation service which relies on a workload estimation model to predict the execution time of a job using application specific parameters, and on a capacity estimation model to estimate the execution time of a submitted job using resource specific parameters.

6. CONCLUSIONS

In this paper, we presented a generic QoS support infrastructure for Grid computing which has been realized in the context of the EU Project GEMSS for the support of time-critical medical simulation applications. In future, we plan to provide QoS support for composite Grid services and to extend our infrastructure towards compliance with WSRF[35].

REFERENCES

1. Apache Tomcat. http://jakarta.apache.org/tomcat/
2. W. Backfrieder, M. Forster, S. Benkner, G. Engelbrecht. Locally Variant VOR in Fully 3D SPECT within A Service Oriented Environment. Proceedings of the International Conference on Mathematics and Engineering Techniques in

Medicine and Biological Sciences, CSREA Press, p. 216-221, Las Vegas, USA, June 2003.

3. S. Benkner, G. Berti, G. Engelbrecht, J. Fingberg, G. Kohring, S. E. Middleton, R. Schmidt. "GEMSS: Grid-infrastructure for Medical Service Provision", HealthGrid 2004, Clermont-Ferrand, France, 2004.
4. S. Benkner, I. Brandic, G. Engelbrecht, R. Schmidt. VGE - A Service-Oriented Grid Environment for On-Demand Supercomputing. Proceedings of the Fifth IEEE/ACM International Workshop on Grid Computing (Grid 2004), Pittsburgh, PA, USA, November 2004.
5. The BioGrid Project. http://www.bio-grid.net/index.jsp
6. BiOpera - Process Support for BioInformatics. ETH Zürich, Dept. of Computer Science. http://www.inf.ethz.ch/personal/bausch/bioopera/main.html
7. BiomedGrid Consortium. http://binfo.ym.edu.tw/grid/index.html
8. R. Buyya. "Economic-based Distributed Resource Management and Scheduling for Grid Computing", PhD Thesis, Monash University, Melbourne, Australia, 2002.
9. J. Cao, F. Zimmermann. "Queue Scheduling and Advance Reservations with COSY", Proceedings of the International Parallel and Distributed Processing Symposium, Santa Fe, New Mexico, USA, 2004.
10. D. M. Jones, J. W. Fenner, G. Berti, F. Kruggel, R. A. Mehrem, W. Backfrieder, R. Moore, A. Geltmeier. "The GEMSS Grid: An evolving HPC Environment for Medical Applications", HealthGrid 2004, Clermont-Ferrand, France, 2004.
11. S. Chokhani, W. Ford, R. Sabett, C. Merrill, S. Wu. Internet X.509 Public Key Infrastructure Certificate Policy and Certification Practices Framework, http://www.ietf.org/rfc/rfc3647.txt, The Internet Society, 2003.
12. I. Foster, C. Kesselman, J. Nick, S. Tuecke. "The Physiology of the Grid: An Open Grid Services Architecture for Distributed Systems Integration", Open Grid Service Infrastructure WG, Global Grid Forum, 22 June 2002.
13. A. Gill, M. Surridge, G. Scielzo, R. Felici, M. Modesti, G. Sardu. RAPT: A Parallel Radiotherapy Treatment Planning Code. In: Liddell H, Colbrook A, Hertzberger B, Sloot P, editors. High Performance Computing and Networking Europe, Springer LNCS 1996. p. 183-193.
14. The GEMSS Project: Grid-Enabled Medical Simulation Services, EU IST Project, IST-2001-37153, http://www.gemss.de/
15. The GRASP Project. http://eu-grasp.net/
16. The Japanese BioGrid Project. http://www.biogrid.jp/
17. S. Ley, D. Mayer, B. Brook, E. van Beek, C. Heusell, R. Hose, D. Rinck, H. Kauczor. Radiological imaging as the basis for a simulation software to advance individualised inhalation therapies. Eur Radiol 2001, 11 (Suppl):216-217.

18. Li JK-J. The Arterial Circulation: Physical Principles and Clinical Applications. Totowa, NJ: Humana Press; 2000.
19. Koch R.M., Roth S.H.M., Gross M.H., Zimmermann A.P., Sailer H.F. A framework for facial surgery simulation. In: Proceedings of the 18th spring conference on Computer graphics; ACM Press; 2002. p. 33-42.
20. The MammoGrid project. http://mammogrid.vitamib.com/
21. The myGrid Project. http://mygrid.man.ac.uk/
22. Maui Cluster Scheduler. http://www.clusterresources.com/products/maui/
23. D. A. Menasce. "QoS-Aware Software Components", Internet Computing Online, Vol. 8, No. 2, p.91-93, 2004.
24. S.E. Middleton, J. Herveg, F. Crazzolara, D. Marvin, Y. Poullet, GEMSS Security and Privacy for a Medical Grid, Methods of Information in Medicine, 2005, to appear.
25. OpenMolGRID - Open Computing GRID for Molecular Science and Engineering. http://www.openmolgrid.org/
26. A. Panagakis, A. Litke, A. Doulamis, N. Doulamis, T. Varvarigou, E. Varvarigos. An Advanced Grid Architecture for a Commercial Grid Infrastructure. The 2nd European Across Grids Conference, Nicosia, Cyprus, January 2004, Springer Verlag.
27. A. Roy, V. Sander. Advance Reservation API, GGF Scheduling Working Group, 2002. http://www.ggf.org/documents/GFD/GFD-E.5.pdf
28. Tittgemeyer M, Wollny G, Kruggel F. Visualising deformation fields computed by non-linear image registration. Computing and Visualization in Science 2002, 5(1):45-51.
29. World Wide Web Consortium. Web Services Architecture, W3C Working Group Note 11 February 2004. http://www.w3.org/TR/2004/NOTE-ws-arch-20040211/
30. SOAP Version 1.2. http://www.w3.org/TR/soap/
31. Vienna Grid Einvironment. http://www.par.univie.ac.at/project/vge/
32. Web Services - Axis. http://ws.apache.org/axis/
33. Web Services Description Language (WSDL) 1.1, http://www.w3.org/TR/wsdl
34. Web Service Level Agreement (WSLA) Language Specification. http://www.research.ibm.com/wsla/WSLASpecV1-20030128.pdf, IBM 2003.
35. OASIS. Web Services Resource Framework (WSRF) Technical Committee. http://www.oasis-open.org/committees/wsrf
36. Web Service Security. SOAP Message Security 1.0, OASIS Standard 200401, March 2004.

LARGE-SCALE SIMULATION AND PREDICTION OF HLA-EPITOPE COMPLEX STRUCTURES

PNG EAK HOCK ADRIAN

Bioinformatics Group, Nanyang Polytechnic
180 Ang Mo Kio Avenue 8, Singapore 569830, Singapore
Email: Adrian_PNG@nyp.gov.sg

TAN TSU SOO

Bioinformatics Group, Nanyang Polytechnic
180 Ang Mo Kio Avenue 8, Singapore 569830, Singapore
Email: TAN_Tsu_Soo@nyp.gov.sg

CHOO KENG WAH

Bioinformatics Group, Nanyang Polytechnic
180 Ang Mo Kio Avenue 8, Singapore 569830, Singapore
Email: CHOO_Keng_Wah@nyp.edu.sg

The predictability of Human Leukocyte Antigens (HLA) binding to their target epitopes has been a holy grail in the field of Immunoinformatics. Currently, predictions are performed using two approaches: machine learning techniques and structural analysis. The latter often requires a large amount of computation in performing molecular simulations and post-process analysis. In this study, our objective is to create a vast number of hypothetical structural variations of HLA-peptide complexes. We present a prototype large-scale molecular modeling system to generate molecular structures that can then be used to determine binding efficiencies of HLA-peptide pairs. Understanding the binding potential of various epitopes allows us to suggest potential vaccine candidates for downstream validation. The system was implemented using Grid Services that embed functionalities of various software tools used for the molecular modeling of protein structures.

1. INTRODUCTION

Major Histocompatibility Complex (MHC) plays a vital role the mammalian immune response. In humans, this highly complex structure is also known as the

Human Leukocyte Antigen (HLA) and as many as 1,972 alleles have been identified.[1]

Associations between the highly polymorphic MHC loci and several human diseases suggest a possible genetic basis of their predisposition.[2,3] Focus has been shifted from a mapping an MHC allele with a particular disease to determining the specific peptides presentation to MHC molecules having clearly defined sequences. Different MHC alleles recognize different peptides and the binding probabilities of peptide ligands to MHC molecules are dynamic.[4] The current challenge is to screen the sequences for candidate MHC ligands for potential T-cell epitopes. Identification of such ligands that are associated with a particular disease, can lead to the development of potential peptide vaccines.[5]

The successful sampling of short peptides from a pool of viral or bacterial protein sequences using MHC-peptide binding prediction programs depends on the accuracy of their algorithms. A number of computational methods have been developed for the prediction of MHC-peptide binding.[6-22] Using data from allele specific binding experiments – sequence binding motif analysis;[6] weight matrices,[7-9] Artificial Neural Networks,[10-12] Hidden Markov Models,[13] and iterative stepwise discriminant meta-algorithm[14] have been applied for predictions. These algorithms have been used to predict peptide binding to very few MHC molecules because binding data is not available for many alleles. Protein threading[15-18] and side-chain packing[19-22] techniques have been applied in molecular mechanics based MHC-peptide binding predictions.

With a high variation of HLA allele types and combinatorial number of possible epitopes, molecular modeling of these complexes can be computationally expensive. Grid computing aims to provide infrastructure to facilitate the sharing of resources. Pooled together, individual machines located disparately, can provide a collaborative platform for staging, execution and analysis. And with a sizable pool of computers, we can hope to achieve high-throughput molecular modeling of MHC-peptide complexes.

2. GMM ARCHITECTURE

The Grid Molecular Modeling (GMM) infrastructure we have designed is a generic platform, combining tools commonly used in molecular modeling. However, for the interest of this publication, we have implemented a specialized framework for the molecular modeling of MHC-peptide complexes. Coordinate files of crystallized MHC-peptide complexes are obtained from Protein Databank (PDB)[23] and sequence information from IMGT database.

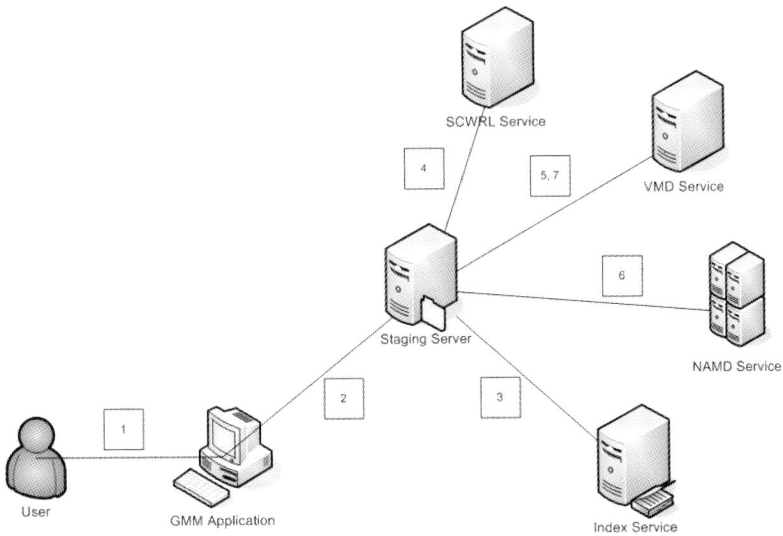

Fig. 1. System architecture of Grid Molecular Modeling System. (1) User configures the workflow on his/her workstation. (2) When the workflow is completed, it is submitted to the staging server. (3) The staging server is central in generating required files, uses the Index Services to identify suitable nodes and sends jobs with corresponding service requirements. (4) A suitable PDB file containing the structure and a file containing the required sequence are transferred via Reliable File Transfer (RFT) to a SCWRL Service node. SCWRL is then executed to generate the new structure file. (5) The new PDB file and VMD script with instructions to prepare the PDB file are then transferred to a VMD Service node. (6) The modified PDB file is then sent to a third node, along with a configuration file for molecular dynamics. We placed the third service on a machine with multiple CPUs (or cluster of computers) as this process usually takes the longest. (7) The resulting PDB file is then redirected to the VMD with post-dynamics instructions. When completed, the final structure file is deposited on the staging server.

The framework consists of three different software: SCWRL,[24] VMD[25] and NAMD.[26] Each tool has a specialized role in this system, which will be discussed further in later sections. The latest production release of Globus Toolkit[27] version 3.2.1 was used as glue to make these services available on the Grid. The overall system architecture is described in Fig. 1.

3. IMPLEMENTATION

The framework was implemented using Globus Toolkit 3.2.1 for infrastructure support. The staging server is the core of the system. It coordinates the sequence to transfer prerequisite files and execute relevant services based on the requirements submitted by the user. Figure 2 illustrates the tasks and order required to produce the target molecular structures.

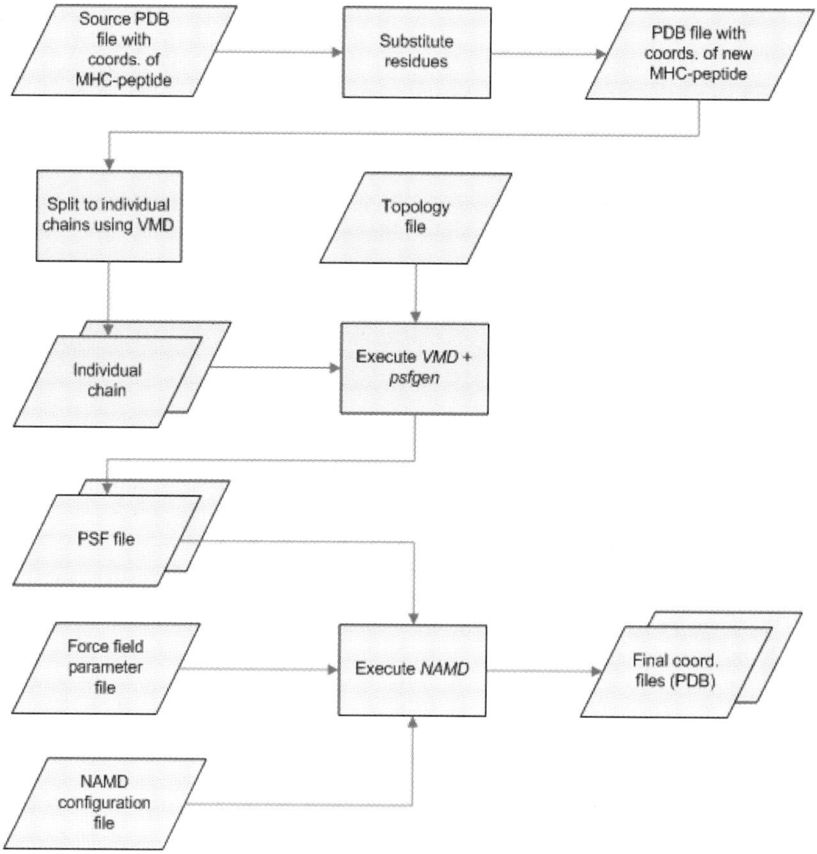

Fig. 2. General workflow for predicting MHC-peptide complex structures.

3.1. SCWRL

SCWRL[24] developed by Dunbrack's group was designed to rapidly and accurately substitute side-chains in existing molecular structure files. SCWRL was used in our implementation to modify MHC-peptide structures archived at the Protein Databank.[23] HLA sequences obtained from IMGT/HLA[1] database and aligned to the HLA sequence from the PDB file. For alignments, the sequences are read from pre-aligned sequences in MSF formats are used. Amino acids that differ are then uppercased and similar amino acids lowercased. For each job sent, both the PDB file and target sequences are sent to the machine hosting the SCWRL service.

3.2. VMD

VMD[25] is a molecular visualization tool that can have scripting support for manipulating molecular structures. In our framework, VMD is used in generating prerequisite files for molecular dynamics executed in NAMD.

The role of VMD is as follows: the PDB file with substituted side-chain configurations by SCWRL is separated into individual chains. Together with a topology file describing atom types and charges, the selected chains are used to generate Protein Structure files (PSF) with the executable `psfgen`. The PSF files are prerequisite for molecular dynamic operations by NAMD. A script to perform these actions is dynamically generated and executed on VMD Service nodes.

3.3. NAMD

The side-chain coordinates determined by SCWRL may result in the overall structure being unstable. Molecular dynamics simulations compute atomic trajectories by solving equations of motion numerically using empirical force fields, such as the CHARMM[28] force field. In our framework, NAMD[26] is responsible for performing molecular dynamics simulations on the new MHC-peptide model. As NAMD was designed for use on parallel machines, we placed the NAMD Service on a small ROCKS[29] cluster.

Output PSF files from VMD Services are then channeled to this node along with the preferred force field parameter file. A configuration file details the molecular dynamics simulation set up is also transported to this node before NAMD is executed.

4. EXAMPLE OF USAGE

For this project, we used a PDB structure representation of HLA B*5301 and a peptide `TPYDINQML` from HIV-2 Gag Protein (PDB ID: 1A1M). The nonamer was separated from the original PDB file and mutated at the 9th position from Leucine to Proline (`TPYDINQMP`) using the SCWRL Service. The VMD Service, with necessary instructions to modify the model then execute `psfgen`, generated the PSF file for the new molecule. The resulting files were then submitted to the NAMD Service with specific requirements embedded in the configuration file. When the process was completed, the modified epitope was placed back with its partnering MHC molecule.

The results of the trial are demonstrated in Fig. 3. The original model obtained from the PDB repository is shown in (A), and the modified complex in (B). It was observed that at least one side-chain was in poor orientation with respect to the

MHC molecule. For example, the side-chain of Tyrosine in position 2 (indicated by the dotted circle) is too close to the beta-sheet that forms the groove floor of the MHC molecule. This suggests that further refinement of the model is necessary to ensure that both molecules are in stable distances from each other.

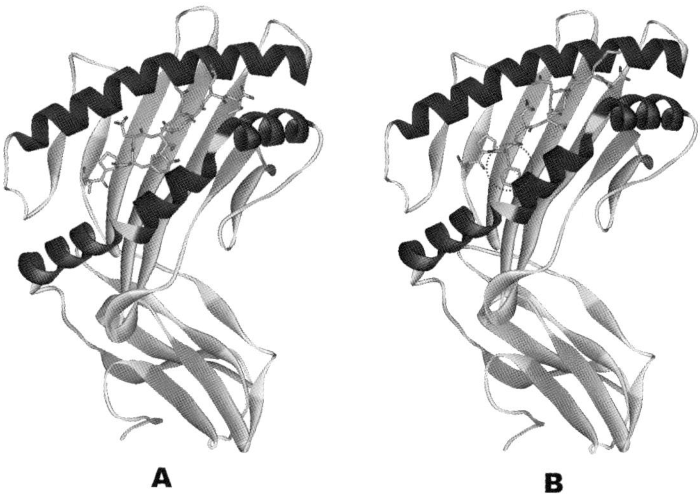

Fig. 3. Molecular structure of the original (A) and variant (B) MHC-peptide.

5. CONCLUSIONS

The prediction of peptide binding to MHC molecules is described as a two-fold problem, the first being protein folding[30] and the second molecular interactions.[31-33] The designed system allows us to achieve the first, and a Grid-based solution allows for large-scale model generation required molecular models. However, further development is necessary to improve the quality of models generated. Such models would then enable us to progress to the next stage, which is calculating the binding affinity by means of determining intermolecular interactions between the MHC molecule and the candidate peptide.

REFERENCES

1. Robinson J, Waller MJ, Parham P, de Groot N, Bontrop R, Kennedy LJ, Stoehr P, Marsh SGE. IMGT/HLA and IMGT/MHC: sequence databases for the study of the major histocompatibility complex. *Nucl Acid Res* 2003; **31**: 311–314.

2. Eckels DD. MHC: Function and implication on vaccine development. *Voc Sang (Suppl 2)* 2000; **78**: 265–267.
3. McDevitt HO. Discovering the role of the major histocompatibility complex in the immune response. *Annu Rev Immunol* 2000; **18**: 1–17.
4. Rammensee HG, Fried T, Stevanoviic S. HC ligands and peptide motifs: first listing. *Immunogenetics* 1995; **41**: 178–228.
5. Buus S. Description and prediction of peptide-MHC binding: The 'Human MHC Project'. *Curr Opin Immunol* 1999; **11**: 209–213.
6. Sette A, Buus S, Appella E, Smith JA, Chesnut R, Miles C, Colon SM, Grey HM. Prediction of major histocompatibility complex binding regions of protein antigens by sequence pattern analysis. *Proc Natl Acad Sci* 1989; 86: 3296–3300.
7. Parker KC, Shields M, DiBrino M, Brooks A, Coligan JE. Peptide binding to MHC class I molecules: implications for antigenic peptide prediction. *Immunol Res* 1995; **14**:34–57.
8. Schafer JR, Jesdale BM, George JA, Kouttab NM, De Groot AS. Prediction of well-conserved HIV-1 ligands using a matrix-based algorithm, EpiMatrix. *Vaccine* 1998; **16**: 1880–1884.
9. Udaka K, Wiesmuller KH, Kienle S, Jung G, Tamamura H, Yamagishi H, Okumura K, Walden P, Suto T, Kawasaki T. An automated prediction of MHC class I-binding peptides based on positional scanning with peptide libraries. *Immunogenetics* 2000; **51**: 816–828.
10. Adams HP, Koziol J. A Prediction of binding to MHC class I molecules. *J Immunol Methods* 1995; **185**: 181–190.
11. Honeyman MC, Brusic V, Stone NL, Harrison LC. Neural network-based prediction of candidate T-cell epitopes. *Nat Biotechnol* 1998; **16**: 966–969.
12. Brusic V, Rudy G, Honeyman G, Hammer J, Harrison L. Prediction of MHC class II-binding peptides using an evolutionary algorithm and artificial neural network. *Bioinformatics* 1998; **14**: 121–130.
13. Mamitsuka H. Predicting peptides that bind to MHC molecules using supervised learning of hidden Markov models. *Proteins: Structure Function and Genetics* 1998; **33**: 460–474.
14. Mallios RR. Predicting class II MHC/peptide multi-level binding with an iterative stepwise discriminant analysis meta-algorithm. *Bioinformatics* 2001; **17**: 942–948.
15. Altuvia Y, Schueler O, Margalit H. Ranking potential binding peptides to MHC molecules by a computational threading approach. *J Mol Biol* 1995; **249**: 244–250.
16. Altuvia Y, Sette A, Sidney J, Southwood S, Margalit H. A structure-based algorithm to predict potential binding peptides to MHC molecules with hydrophobic binding pockets. *Hum Immunol* 1997; **58**: 1–11.

17. Schueler-Furman O, Elber R, Margalit H. Knowledge-based structure prediction of MHC class I bound peptides: a study of 23 complexes. *Fold Des* 1998; **3**: 549–564.
18. Schueler-Furman O, Altuvia Y, Sette A, Margalit H. Structure-based prediction of binding peptides to MHC class I molecules: application to a broad range of MHC alleles. *Protein Sci* 2000; **9**: 1838–1846.
19. Rognan D, Lauemoller SL, Holm A, Buus S, Tschinke V. Predicting binding affinities of protein ligands from three-dimensional models: application to peptide binding to class I major histocompatibility proteins. *J Med Chem* 1999; **42**: 4650–4658.
20. Lee C, McConnell HM. A general model of invariant chain association with class II major histocompatibility complex proteins. *Proc Natl Acad Sci* 1995; **92**: 8269–8273.
21. Kangueane P, Sakharkar MK, Lim KS, Hao H, Lin K, Ren EC, Kolatkar PR. Knowledge based grouping of modeled HLA peptide complexes. *Hum Immunol* 2000; **61**: 460–466.
22. Doytchinova IA, Flower DR. Toward the quantitative prediction of t-cell epitopes: comfa and comsia studies of peptides with affinity for the class I MHC molecule hla-a0201. *J Med Chem* 2001; **44**: 3572–3581.
23. Bernstein FC, Koetzle TF, Williams GJB, Meyer Jr EF, Brice MD, Rodgers JR, Kennard O, Shimanouchi T, Tasumi M. The Protein Data Bank: a computer-based archival file for macromolecular structures. *J Mol Biol* 1977; **112**: 535–542.
24. Canutescu AA, Shelenkov AA, Dunbrack RL. A graph theory algorithm for protein side-chain prediction. *Prot Sci* 2003; **12**: 2001–2014.
25. Humphrey W, Dalke A, Schulten K. VMD - Visual Molecular Dynamics. *J Mol Graphics*, 1996; **14**: 33–38.
26. Kalé L, Skeel R, Bhandarkar M, Brunner R, Gursoy A, Krawetz N, Phillips J, Shinozaki A, Varadarajan K, Schulten K: NAMD2. Greater scalability for parallel molecular dynamics. *J Comp Phys* 1999; **151**: 283–312.
27. Globus – http://www.globus.org/
28. Brooks BR, Bruccoleri RE, Olafson BD, States DJ, Swaminathan S, Karplus M. CHARMM: a program for macromolecular energy, minimization, and dynamics calculations. *J Comp Chem* 1983; **4(2)**: 187–217.
29. Rocks Clusters - http://www.rocksclusters.org/
30. Levitt M, Gerstein M, Huang E, Subbiah S, Tsai J. Protein folding: the endgame. *Annu Rev Biochem* 1997; **66**: 549–579.
31. Jones S, Thornton JM. Principles of protein-protein interactions. *Proc Natl Acad Sci* 1996; **93**: 13–20.
32. Jones S, Marin A, Thornton JM. Protein domain interfaces: characterization and comparison with oligomeric protein interfaces. *Protein Eng* 2000; **13**: 77–82.
33. Conte LL, Chothia C, Janin J. The atomic structure of protein-protein recognition sites. *J Mol Biol* 1999; **285**: 2177–2198.

CONSTRUCTION OF COMPLEX NETWORKS USING MEGA PROCESS GA AND GRID MP

YOSHIKO HANADA

Graduate School, Dept. of Knowledge Engineering & Computer Sciences, Doshisha University, 610-0321 Kyoto, Japan
Email: hanada@mikilab.doshisha.ac.jp

TOMOYUKI HIROYASU AND MITSUNORI MIKI

Dept. of Knowledge Engineering & Computer Sciences, Doshisha University, 610-0321 Kyoto, Japan
Email: {tomo@is, mmiki}@mail.doshisha.ac.jp

In this study, a new Genetic Algorithm for large-scale computer systems comprised of massive processors, named Mega Process GA, is introduced. Our method has a GA-specific database possesses information of searched space. In addition, the local search for non-searched spaces is applied using individuals stored in the database. Applying this local search, the searched space can be expanded linearly in accordance with the increase in computing resources and the exhaustive search is guaranteed under infinite computations. This method was applied to the problem regarding the construction of complex networks for the base study of interactions among proteins. Through the experiments, we indicated the prospect of construction of complex networks by our method. We examine the performance of the proposed method on a distributed computing environment, which is built using the commercially available middleware Grid MP produced by United Devices Inc.

1. INTRODUCTION

Genetic Algorithms (GAs) are among the most effective approximation algorithms for optimization problems. GAs are well suited to parallel processing environments due to their abilities to search with multiple points and their tolerances for extinction of search points. Consequently GAs have found applications in large-scale computing [1-4]. Due to the recent emergence of super PC clusters and Grid computation environments, such as PC Grid comprised of desktop machines for home use or offices, the number of available computational calculation resources is increasing. Therefore, GA that uses large-scale computer systems comprised of massive processors, i.e., Mega Processors, has become feasible. However, the application of GAs on large-scale computing environment, such as grid computation environments, has the drawback that these algorithms lack scalability in their performance improvement in accordance with increases in available computing resources. Therefore, huge computing resources cannot be used effectively. This is caused by overlapping searches.

In this study, a GA for large-scale computing system that has mechanisms to use massive computation resources laconically and to search effectively, the Mega Process GA, is introduced. Our method has a GA-specific database that possesses information regarding the space that has been searched already to avoid overlapping searches and make effective use of computing resources in consideration of scalability of search performance when GAs use large computer resources. In our previous work for Mega Process GA, we proposed the expression of the searched region using schemata and a local search mechanism as a compression method to store large regions searched by several individuals. We then showed that the searched space could be expanded as the number of computing resources increased, enhancing accuracy and reliability by applying the database with the local search mechanism to a GA, and it was clear that the proposed method also showed superior performance with limited computing costs [5]. Nevertheless, in this work, there was the drawback that computing costs increased exponentially in accordance with generations. We proposed a new database and local search for Mega Process GA based on our previous work. It was shown in preparative experiments that our method ensured an effective exhaustive search and had almost the same performance as a conventional GA in primitive functions and test functions of continuous optimization problems [6].

In this paper, the proposed GA was applied to the problem regarding the construction of complex network for the base study of interactions among proteins. We examined the performance of the proposed method on a distributed computing environment composed of machines belonging to Doshisha University and RIKEN Genomic Sciences Center, which was built up by the commercially available middleware produced by United Devices Inc., named Grid MP.[a]

2. CONCEPT OF MEGA PROCESS GA

In this section, we introduce the Mega Process GA. Our method has GA-specific database that carries information regarding the regions that have already been

[a]United Devices : http://www.ud.com

searched to avoid search overlapping. At the same time, the proposed GA performs the local search for the space that is not searched to expand the searched space. To obtain optima earlier, our method of searches uses mainly schemes of GA; any methods of operators such as a crossover and a mutation or a generation alternation model can be applied. To use idle computing resources of enormous computing environments effectively, a local search is applied. The outline of our method is shown in Figure 1.

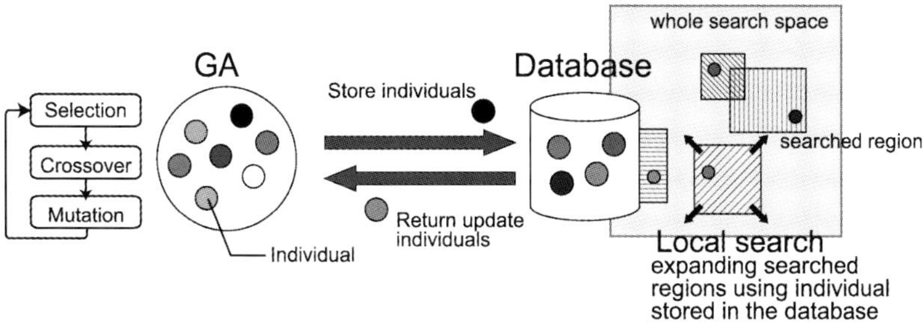

Figure 1. The outline of our method.

2.1. Database

When a database stores all the individual information, it takes a large time to check an individual that has already been searched due to the vast amounts of data. Therefore, all search regions should be stored as highly compressed expressions and checking of individuals stored in the database should not be time-consuming. We used binary-coded individuals, and an individual or set of individuals is represented by 2-dimensional coordinates using the mapping method proposed by Collins, which converts the multidimensional search space to a 2-dimensional plane [7]. In this mapping method, an individual is treated as the coordinates (x, y), where the integer x is coded by Gray-coding from the string composed of genes of $2k$-th loci extracted from the chromosome, and integer y is coded from the string of extracted $(2k+1)$-th loci. Each individual has a one to one correspondence with a set of coordinates. Hence, the 2-dimensional plane can express the whole space. In our proposed database, the set of individuals that have contiguous coordinates on the 2-dimensional plane are represented by a rectangle and are stored by two diagonal vertexes (x_{min}, y_{min}), (x_{max}, y_{max}) of the rectangle. In addition, the best individual in the searched region is stored.

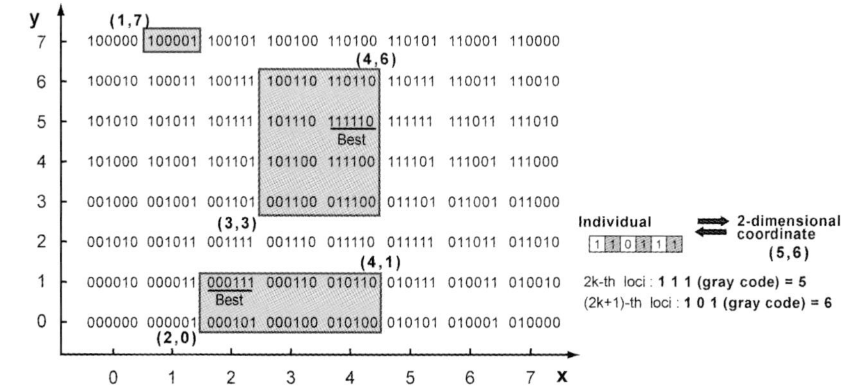

Figure 2. An example of a 6-bit Problem Space created by the Mapping.

	Searched region	Best individual	Size of searched region
data1	(3,3)-(4,6)	111110	8 (4x2)
data2	(2,0)-(4,1)	000111	6 (2x3)
data3	(1,7)-(1,7)	100001	1 (1x1)

Figure 3. Database representing searched regions illustrated in Figure 2.

Figure 2 illustrates the whole search space of a problem consisting of 6 bits represented as a 2-dimensional plane. Examples for a set of individuals represented as rectangles are shown in this figure. Figure 3 shows the database that possesses searched regions shown in Figure 2. In this figure, a rectangle is represented with (x_{min}, y_{min})-(x_{max}, y_{max}), which are diagonal vertexes. This notation enables us to comprehend the quantitative rate of the searched region by calculating the square measure of the rectangle.

2.2. Local Search

Our proposed database possesses searched individuals in the GA search using the expression described in the previous section. In addition, using individuals stored in the database, a local search for the space that has not been searched is applied to use idle computing resources of enormous computing environments effectively.

In our proposed local search, each rectangle stored in the database as searched regions is expanded vertically and horizontally on a 2-dimensional plane. In one step in our proposed local search, each rectangle is expanded vertically and horizontally by 1. Figure 4 illustrates our proposed local search. In this figure, the rectangular regions, or individuals, painted in gray are searched region. The regions surrounded with bold lines are searched additionally in one step of local search. As a result, searched regions are expanded. The directions of longitudinal expansion and lateral expansion are determined by fitness of the best individuals on the edges of the rectangle.

In this study, our method is applied to the problem regarding the construction of complex networks for the base study of interactions among proteins. We describe about complex networks and how to construct them using GAs in the next section.

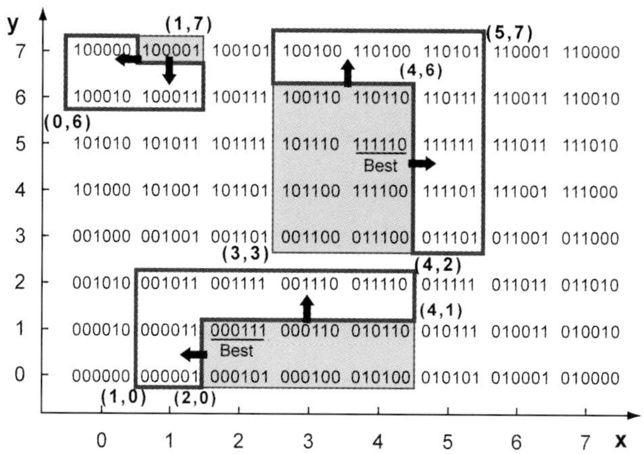

Figure 4. An example of our proposed local search.

3. COMPLEX NETWORKS

In this paper, we applied our method to the problem regarding the construction of complex networks. Complex network is the network that consists of multiple elements in population, or system, and indicates a unique behavior that surpasses total of partial behavior as the whole due to influences and interaction among its elements [8]. Complex networks have been often found in interaction among proteins as shown in Figure5; however, their structures and characteristics have not been clear [9]. Various studies to analyze existing networks have been discussed.

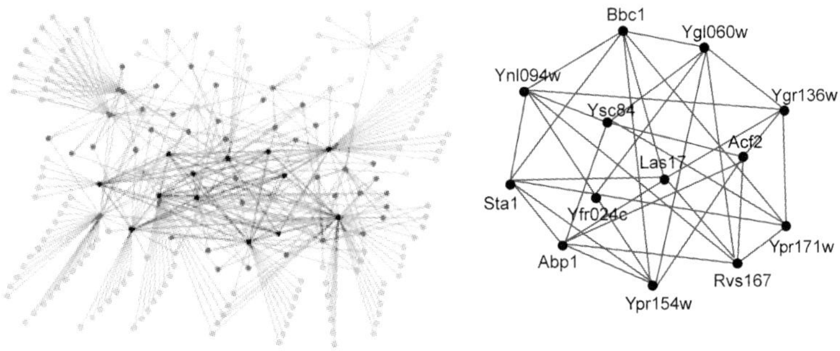

Figure 5. An example for protein-protein interactions.

We apply another approach that focuses on the characteristic in complex network and constructs the network using optimization algorithm. Characteristics of the constructed network are then examined in this approach.

3.1. Model of Problem

Our approach is to construct complex networks by GA that optimizes a representative value of the network. We used the average of curtate distance between nodes that is an objective function that should be minimized and we fixed the number of links and nodes as constrained conditions.

The problem used in here is defined as (1), where D is the average of curtate distance between nodes and D_{ij} is curtate distance between nodes i and j.

$$D = \frac{1}{n(n-1)} \sum_{i=1}^{n} \sum_{j=1}^{n} (D_{ij}) \quad (i \neq j) \tag{1}$$

In this paper, we examine whether networks obtained by GAs minimizing D can satisfy conditions of complex networks. Most existing networks are known in generally as scale-free networks [10]. Scale-free networks are composed of the node named Hub that has huge number of links and the node that are linked with few nodes [9]. These networks have larger cluster coefficients and smaller average distances between nodes than random networks [11]. Cluster coefficient is one of characteristics in network and defined as (2) where C is the cluster coefficient and C_i is the cluster coefficient at node i [12].

$$C \equiv \frac{1}{N}\sum_{i=1}^{N} C_i,\ C_i = \frac{2Ei}{k_i(k_i - 1)} \tag{2}$$

By comparing obtained networks by GA with random networks, we acknowledge that obtained networks by GAs have the characteristic of scale-free networks, i.e., complex networks, if they have larger cluster coefficient and smaller average distances than those of random networks.

3.2. Construction of Complex Networks by GA

3.2.1. *Coding Method*

Each topology of nodes has to be coded into chromosomes to optimize complex networks by GA. First, topologies are represented by matrix, each element a_{ij} of which indicates connection between node n_i and node n_j ($0 <= i, j < N$: N is the number of nodes). a_{ij} stands at '1' where node n_i and node n_j are connected by the link and stands at '1'. '0' indicates that two nodes are not connected. This matrix is symmetric; therefore, we use elements belonging in the upper triangular matrix. Figure 6 shows the genotype of a network composed of 6 nodes and 5 links. In this coding method, the length of chromosome is $N(N\text{-}1)/2$ and the scale of the whole search space is $2^{N(N\text{-}1)/2}$.

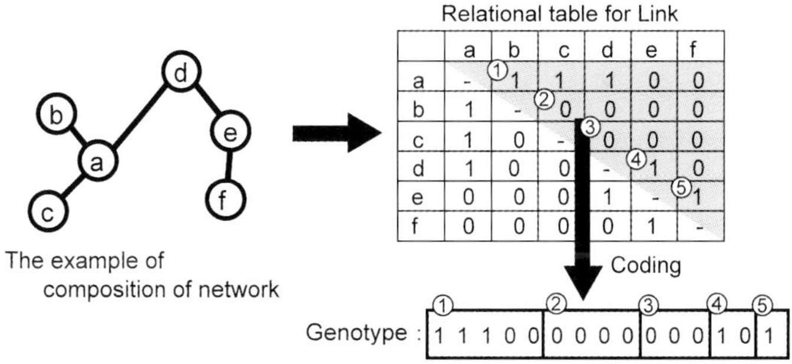

Figure 6. An example for the coding for a network.

3.2.2. *Crossover*

In this problem, there are two constraints: The total number of links stays constant and each node has to be connected directly or remotely to other nodes. Networks shown in Figure 7 are not feasible.

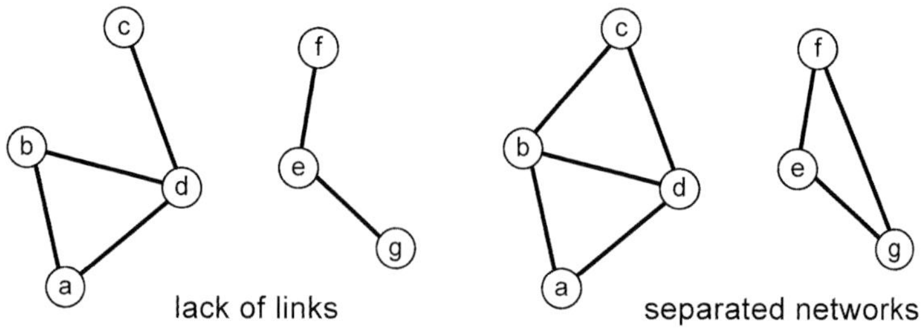

Figure 7. Examples of non-feasible networks in 7nodes-8links problem.

To avoid generating of non-feasible networks, we introduce the crossover that keeps the number of links. In our crossover, different links between two networks are exchanged in random. Figure 8 shows an instance for this crossover.

Modification is then applied if the generated network does not satisfy the second constrained condition as shown in Figure 9. In this figure, one of loose-fitting links is cut and two separate networks are then bridged by the arbitrary link.

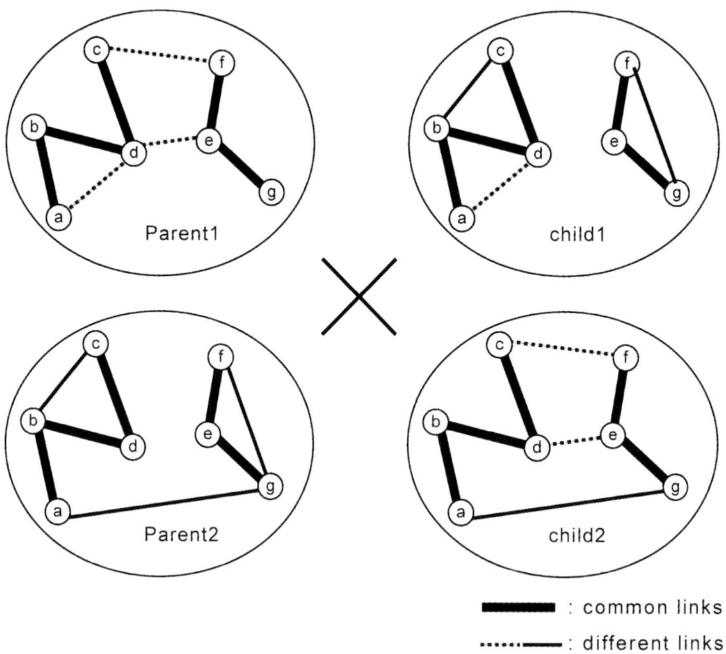

Figure 8. Examples of Crossover.

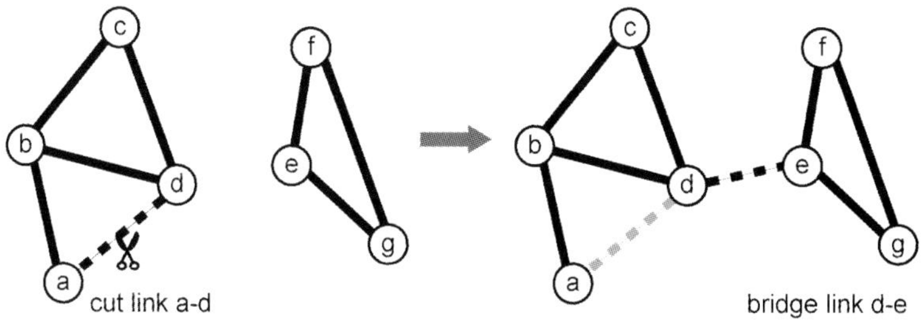

Figure 9. Modification of a non-feasible network.

These crossover methods were applied to construct networks. We then examined whether networks obtained by GA and our proposed GA minimizing the average of curtate distance between nodes can satisfy conditions of complex networks.

4. NUMERICAL EXPERIMENT

It was shown in preparative experiments that our method ensured an effective exhaustive search and had almost the same performance as a conventional GA in primitive functions and test functions of continuous optimization problems [6]. Here, its effectiveness in limited computation costs was examined in the network construction problems by comparing it to a conventional GA. We then examined whether networks obtained by GA and our proposed method can satisfy conditions of complex networks. In addition, we performed proposed method on a distributed computing environment built up by Grid MP and showed that our method could expand the searched space in accordance with the increase in computing resources.

4.1. Performance with Limited Computation Cost

In our method, there are various usages for the proposed database with GA. In the experiment that used one computation node, a local search was conducted alternately with a GA using individuals stored in the database. To obtain a good solution at as early a stage as possible, the search was advanced based on the GA. The proposed method is outlined in Figure 10.

First, the population in the GA is initialized. The number of individuals in the database at initial generation is 0. Next, genetic operations, such as crossover, mutation, and selection, are conducted in the population of GA, and the best individual of the population is preserved in the database. However, if this individual is already preserved in the searched region of the database, it will not be preserved again. Next, a local search operation is applied to all individuals in the database. When an individual with better fitness than the individual of interest is found, the copy of the individual will replace the worst individual in the GA population.

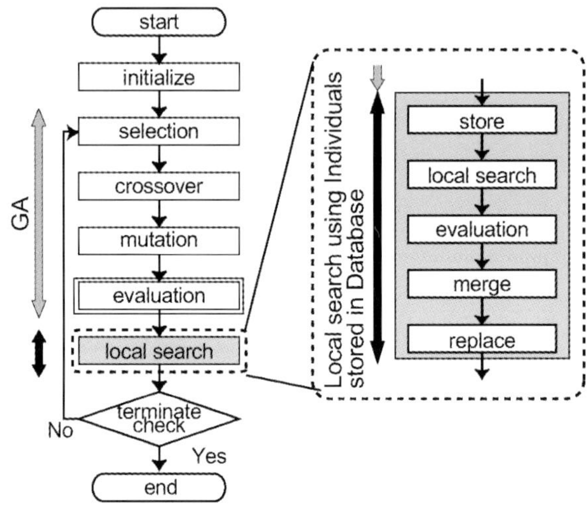

Figure 10. The flow of the proposed GA performed in one computation node.

The proposed GA and a conventional GA are compared in the limited number of evaluations. In this experiment, we took cases of the alignment of node of benchmarks problem used in Traveling Salesman Problem; three instances, (29 nodes, 45 links), (51 nodes, 76 links) and (101 nodes, 150 links), were used. (*N* node, *M* links) denotes that the number of nodes is *N* and number of links is *M*.

The ER model [13] was used as the alternation in each generation. The population size was 100 and each couple, or parents, generated 20 children by crossover. The maximum numbers of evaluation calculations was limited to $4x10^4$ in (29 nodes, 45 links) and (51 nodes, 76 links). In (101 nodes, 150 links), $8x10^4$ evaluations were achieved.

Table 1 shows the averages of curtate distance between nodes of obtained network by GA and our method in three instances. Cluster coefficients are shown in Table 2 for comparing them with those of random networks.

Table 1. Averages of curtate distance between nodes.

	Conventional GA		Proposed Method		Random Network
	best	average	best	average	Best
29 nodes 45 links	1016.80	1019.81	1015.18	1019.27	1796.93
51 nodes 76 links	38.96	39.02	38.97	39.02	81.30
101 nodes 150 links	38.65	38.75	38.61	38.78	98.58

Table 2. Averages of Cluster coefficients.

	Conventional GA	Proposed Method	Random Network
29 nodes 45 links	0.235	0.229	0.100
51 nodes 76 links	0.125	0.128	0.053
101 nodes 150 links	0.092	0.097	0.035

These results are the best and average fitness obtained by two methods. They are obtained from 20 trials and the result of random network is from 100 trials.

The result shows that our method retains the performance of a conventional GA in limited computational cost though a large amount of evaluation calculation is necessary in the local search. In addition, by comparing an obtained network with random networks, it was clear that obtained network had larger cluster coefficient and smaller average distances than those of random network. Scale-free network is the network that has larger cluster coefficients and smaller average distances between nodes than random networks; therefore we conclude that both methods are effective for constructing scale-free networks, i.e., complex networks.

4.2. Implementation on Grid MP

To discuss effectiveness of our approach on a large-scale computing system, we constructed a distributed computing environment using Grid MP produced by United Devices Inc. as a grid middleware. Parallelization is applicable to this local search by allotting each rectangle, i.e., a searched region stored in the database, to computation nodes. There are no strong dependencies and little communication among searches; therefore, to execute the local search on distributed computing systems is expected to yield high throughput computation.

In the previous section, we have described an implementation of our approach that conducts the local search alternately with a GA in limited computation costs. This implementation is not appropriate on distributed computing environment because of large overhead in synchronization between GAs and local searches. In distributed environment built up with Grid MP, a GA and local search are conducted in asynchronous way shown in Figure 11. The best individual found in GA is stored into the database every generation. The database is kept by the server named MP Server, and the local search that expands the searched region using information stored in the database is executed respectively in worker nodes named Devices, of the system. Each node returns periodically its current searched region to server to update the database. GA refers to the database for information about searched regions during evaluations.

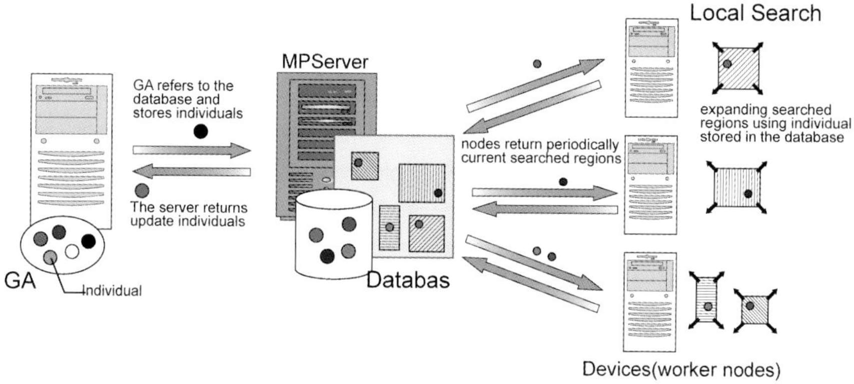

Figure 11. Implementation on Grid MP.

We examined whether it assured the scalability of search performances, i.e., the scale of searched region, against the number of computing resources. We used Grid MP version 4.1. The specification of machines used for the experiments is shown in Table 3. The term "Local Machine" in Table 3 indicates the machine that performs GA and submits jobs regarding local search to the MP Server.

Table 3. Specification of machines used for the experiment.

	Affiliation	#Nodes	Processors	Memory
Local Machine	- -	1	Pentium M 1.8GHz	1GB
MP Server	Doshisha	1	Xeon 2.8GHz x 2	2GB
Devices	Doshisha	9	Xeon 2.4GHz x 2	1GB
Devices	RIKEN GSC	21	Celeron 1.3GHz	896MB

We used 10, 20 and 30 Devices shown in Table 3. Sets of Devices used in this experiment were (3,7), (6,14), and (9,21) where the notation (n_1, n_2) indicates that n_1 nodes of Doshisha and n_2 nodes of RIKEN GSC are used. The parameters of GA were the same as in the previous section except that population size was 150, and the 51 nodes 76 edges problem were examined.

In the local search run on Grid MP, each Device returned the current searched region every 3 minutes. Figure 12 shows the scale of searched region obtained by 2 hours of execution using 10, 20, and 30 Devices shown in Table 3. From this result, it is obvious that the searched space can be expanded linearly in accordance with the increase in computing resources as we proposed. In the optimization process, the scalability of the searched region against the number of calculation resources is very important. The environment used in this study is a small environment and we should examine the performance of the proposed method on large scale distributed computing environments. This examination will be conducted in our future studies.

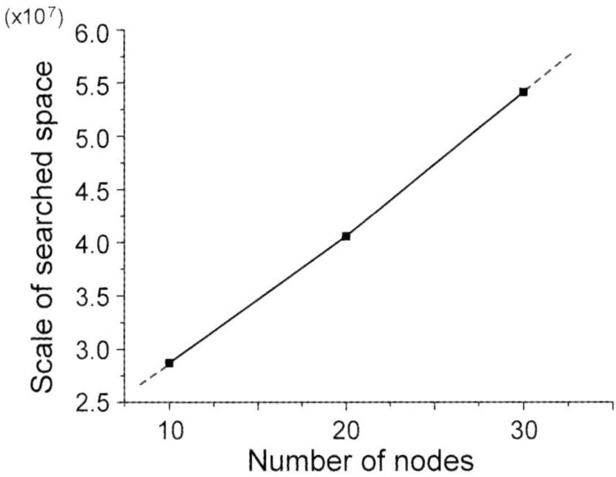

Figure 12. Increase in scale of searched region against the number of nodes.

5. CONCLUSION

GAs are well suited to parallel processing environments in a certain amount of calculation nodes. Due to the recent emergence of super PC clusters and Grid computation environments the number of available computational calculation resources is increasing. Therefore, GA that uses large-scale computer systems comprised of massive processors, i.e., Mega Processors, has become feasible. Mechanisms to use massive computation resources laconically and to assure the scalability of search performances against the number of computing resources are necessary if large-scale computer systems are available.

In this paper, a GA for large-scale computing system with mechanisms to use massive computation resources and search effectively, called Mega Process GA, was introduced. Our method has a GA-specific database that possesses information regarding the space that has been searched already to avoid overlapping searches and make effective use of computing resources in consideration of scalability of search performance when GAs use large computer resources. In addition, the local search for non-searched spaces is applied using individuals stored in the database. Applying this local search, the searched space can be expanded linearly in accordance with the increase in computing resources.

Our proposed method was applied to the problem regarding the construction of complex network for the base study of interactions among proteins. In limited computational cost, it was clear that our method retained superior performance of a conventional GA though a large amount of evaluation calculation was necessary in the local search. We then performed proposed method on a distributed computing environment built up by Grid MP and showed that our method could expand the searched space in accordance with the increase in computing resources. The environment used in this study is a small environment. In network construction problem composed of huge nodes and links, enormous amount of computation are required due to its wide spread search space. In consequence, we should examine the performance of the proposed method on large scale distributed computing environments. This examination will be conducted in our future studies.

ACKNOWLEDGMENTS

We are grateful to Akihiko Konagaya and Fumikazu Konishi of RIKEN Genomic Sciences Center as well as Hiroyuki Kobayashi of Sumisho Electronics Co., Ltd. for valuable discussion and contributions to the development of the distributed computing environment built using Grid MP.

REFERENCES

1. Y. Tanimura et al.: Development of Master-Worker System for The Computational Grid. Information Processing Society of Japan: Computing System. Vol 45, No. SIG6 (ACS6), pp. 197-207, May 2004. (in Japanese)

2. H. Imade et al.: A Grid-Oriented Genetic Algorithm for Estimating Genetic Networks by S-Systems, Proc. SICE Annual Conference, pp. 3317-3322, 2003.
3. H. Imade et al.: A framework of grid-oriented genetic algorithms for large-scale optimization in bioinformatics.
4. H. Nakata et al.: Protein structure optimization using Genetic Algorithm on Jojo Journal of Information Processing Society of Japan. 2002-HPC-93, pp. 155-160, 2003. (in Japanese)
5. Y. Hanada et al.: Mega Process Genetic Algorithm Using Grid MP, Life Science Grid 2004, LNAI, vol. 3370, pp. 152-170, 2004.
6. Y. Hanada et al.: An Improvement of Database with Local Search Mechanisms for Genetic Algorithms in Large-Scale Computing Environments, 2005 IEEE Congress on Evolutionary Computation. (to appear)
7. T. Collins: Understanding Evolutionary Computing: A Hands-on Approach, KMI-TR-48. September 1997.
8. G. Nicolis et al.: Exploring complexity. R. Piper GmbH and Co. KG Verlag, 1989.
9. A.-L. Barabasi: Linked - the new science of network, Perseus Books, 2002.
10. A.-L. Barabasi and R. Albert: Emergence of scaling in random networks. Science 286, pp. 509-512, 1999.
11. P. Erdoas et al.: Publication of Mathematics Institute, Hungary Academy of Science 5, 17, 1960.
12. W. Souma et al.: The role of small world network. Technical report of IEICE, NGN2001-12. (in Japanese)
13. D. Thierens et al.: Elitist Recombination: an integrated selection recombination GA, Proceedings of 1st IEEE Conference on Evolutionary Computation, pp. 508-512, 1994.

ADAPTING THE PERCEPTRON FOR NON-LINEAR PROBLEMS IN PROTEIN CLASSIFICATION

MARTIN CHEW WOOI KEAT, ROSNI ABDULLAH, ROSALINA ABDUL SALAM

Faculty of Computer Science, Universiti Sains Malaysia
Georgetown, Penang, Malaysia
Email: martin.chew@motorola.com ,{ rosni, rosalina }@cs.usm.my

Perceptrons are simple, yet effective classifiers for linearly separable problem domains. For non-linearly separable problems, back-propagation networks are typically used. However, back-propagation networks require greater effort to implement and parallelize, compared to the simple perceptron. In order to maintain simplicity, as well as the ability to cope with non-linearly separable problems, we explored the use of multi-perceptrons. A multi-perceptron architecture is built by interconnecting simple perceptrons. Based on an example of how a multi-perceptron can be used to cope with the XOR problem, we adapted the example for the purpose of protein classification. We then investigated the parallelization of the perceptron in a protein classification context. Implementation considerations are discussed. Our eventual goal is an array of parallelized multi-perceptrons for the purpose of protein classification.

1.1. Introduction

This concept paper relates to the field of protein classification. Neural networks [5, 6, 7] and neural network arrays [1, 2] have been used before in the field of protein classification. Two known examples are based on the simple perceptron and an adapted form of the weightless neural network. Since the weightless array requires only a single pass on the sample data, it is recommended to be used in a "rapid proto-typing" manner, to investigate the potential of new protein encoding schemes (2-gram, 3-gram, etc.). For a more thorough classification, the perceptron array is recommended. However, perceptrons are effective for linearly-separable problems only, with non-linearly separable problems being delegated to back-propagation networks. Back-propagation networks are more complex, and hence, require more effort to implement and parallelize compared to perceptrons. This paper investigates how a perceptron may be extended to cope with non-linear problems, and adapted for the purpose of protein classification. A multi-perceptron architecture maintains the basic simplicity of the perceptron (i.e., the delta learning rule), but with the essential addition of the capability to deal with non-linearly separable problems.

1.2. Multi-Perceptron Concept

The structure of a single perceptron is very simple. There are multiple inputs, a bias and an output. Each of the inputs and the bias is connected to the main perceptron by a weight. A weight is generally a real number between 0 and 1. When the input number is fed into the perceptron, it is multiplied by the corresponding weight. After this, the weights are all summed up and fed through a hard-limiter. Basically, a hard-limiter is a function that defines the threshold values for firing the perceptron. The way a perceptron learns to distinguish patterns is through modifying its weights. The learning is accomplished by adjusting the weights by the difference between the desired output and the actual output. Learning on a perceptron is guaranteed, as stated by the Perceptron Convergence Theorem which states that if a solution can be implemented on a perceptron, the learning rule will find the solution in a finite number of steps [3].

Perceptrons can only solve problems where the solutions can be divided by a line (or hyperplane) - this is called linear separation. An example of a linearly separable problem is the OR problem. The representation of the problem is given in Table 1. When translated to a graph (refer to Figure 1), TRUE outputs (the symbol

Table 1. OR Problem

X1	X2	Output	Symbol
1	0	1	X
0	1	1	X
1	1	1	X
0	0	0	O

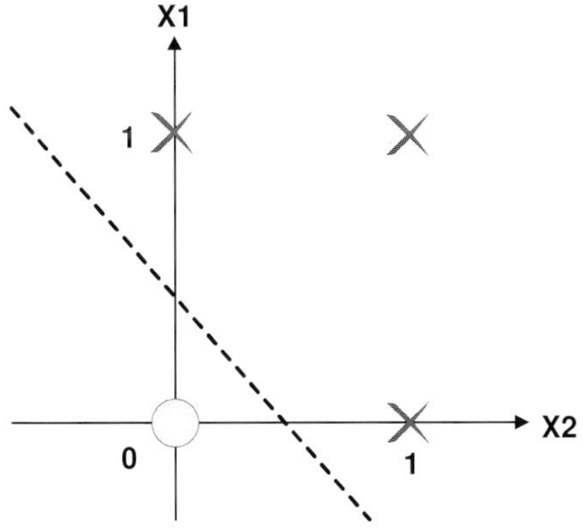

Figure 1. OR problem with the TRUE and FALSE classes linearly separated by a straight line.

"X") form a class, and FALSE outputs (the symbol "O") form another class. One class is linearly separable from the other.

The perceptron is ineffective when it is unable to correctly draw a line that divides two groups of points. For example, a perceptron cannot solve the XOR problem (refer to Table 2 and Figure 2).

Table 2. XOR Problem

X1	X2	Output	Symbol
1	0	1	X
0	1	1	X
1	1	0	O
0	0	0	O

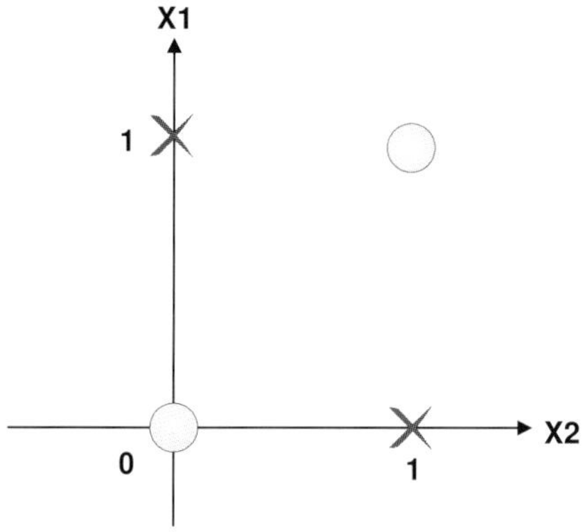

Figure 2. XOR problem with the TRUE and FALSE classes unable to be linearly separated.

However, the XOR problem can be solved by using three perceptrons. The key lies in splitting up the XOR problem into three different parts. For example (assuming x1 and x2 are the inputs and y is the output):

y = (x1 and (not x2)) or ((not x1) and x2)
y = (x1 or x2) and (not(x1 and x2))
therefore:
y1 = x1 or x2
y2 = not (x1 and x2)
y = y1 and y2

The problem is now broken down into three different linearly separable problems. The results from the first two equations will be fed into a third linear separable problem. The multi-perceptron architecture representing the interconnection between the three linearly separable problems is given in Figure 3.

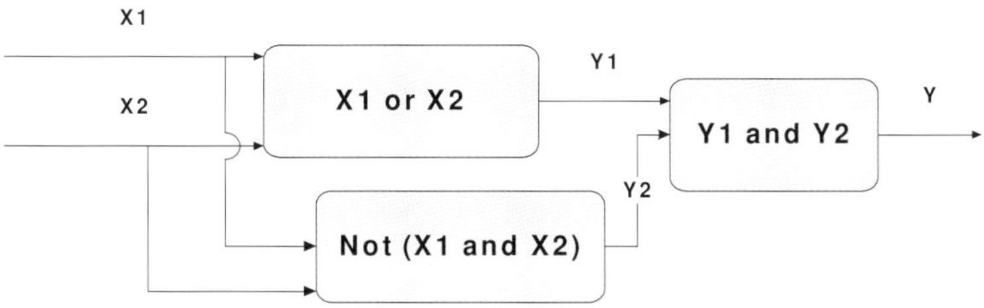

Figure 3. Multi-perceptron architecture.

1.3. Multi-Perceptron Concept Adapted for Protein Classification

Protein classification is defined as the identification of the protein family to which a given unknown protein sequence belongs. A protein family is a collection of protein sequences which share structural or functional similarities. A protein sequence has to be encoded into an array of real values before it is able to be processed by a neural network. There are various possible encoding schemes (2-gram, 3-gram, etc.) and the encoding scheme which best abstracts the unique features of a particular protein family should be used to encode the member sequences of that family.

For the purpose of this paper, we used 2-gram encoding. When 2-gram encoding is applied on the sequences of a particular protein family, the end results is a cumulative frequency count for each pairing (AA, AB, AC, etc.). Since there are 20 unique amino acids relevant to the human body, we would have a 2-gram population size of 400 pairings (20x20).

If we are using a single perceptron, then our single perceptron will have 400 inputs and a single output. When an unknown protein sequence is submitted to the trained perceptron, and the output is above a certain threshold, then that unknown protein sequence is considered to be from the family abstracted by that particular perceptron.

However, we are implicitly assuming linear separability. In the event this assumption falls apart during the validation of the system, our next step would be to utilize a multi-perceptron architecture.

By observing the cumulative frequency count data, we would be able to ascertain distinguishing peaks in the frequency distribution. We will then split up the domain into different parts, by having different perceptrons represent different peaks. The outputs from these perceptrons will be fed into a master perceptron which will give the final output (refer to Figure 4).

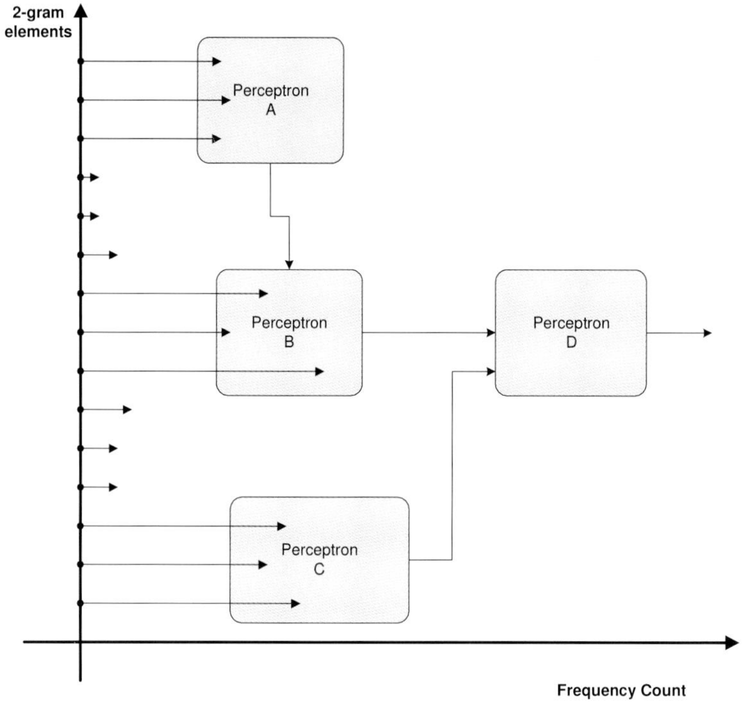

Figure 4. Multi-perceptron concept applied to protein classification.

The sample architecture above allows us to implement AND/OR checks with respect to certain distinguishing peaks in the data. As a result of this, we would have the flexibility to be able to cater to non-linear situations.

1.4. The Benefits of Keeping it Simple

A perceptron is very straightforward to implement, since its structure and learning algorithm is relatively simple. A perceptron object only has a limited number of methods (e.g., read inputs, summation, applying the threshold to determine the output, determining the error, and correcting the weights). Yet, it is powerful enough to cope with numerous classification problems. When we parallelize the perceptron, the basic methods described above are sufficient. We need to only divide up the inputs among the different perceptron instances, for the mathwork to be done in parallel. The master perceptron will gather the sub-totals, and give the final output based on the hard-limiter function used. Error correction could also be done in parallel, all that is needed is the delta value to be sent to the various instances. This simplicity would not be applicable for the case of a parallel back-propagation network, yet a multi-perceptron has the non-linear capability of a back-propagation network. Simplicity is very important because it reduces the performance and implementation overheads of the system.

We parallelized the perceptron using MPI. The basic principle of MPI is that many programs run in parallel, and communicate with each other via messages to coordinate their computations. Each program runs on a node. A node is typically associated with one processor and its memory. Therefore, MPI is said to have a distributed memory architecture. It is often necessary to rewrite large parts of a sequential program before it can use the MPI API. An MPI program is often difficult to debug, since it is hard to modularize a MPI program due to the strong couplings involved between the message sender and receiver [4]. With the simple perceptron, we will be able to keep the use of MPI to a minimum. A multi-perceptron system could then be built on top of the basic perceptron.

Table 3. Basic perceptron array results.

Sample	F1	F2	F3	F4	True?
F1-1	***0.974335***	0.315582	0.495418	0.678569	Yes
F1-2	***0.970818***	0.313173	0.495083	0.683184	Yes
F1-3	***0.950345***	0.281784	0.492670	0.658418	Yes
F2-1	0.580057	***0.889027***	0.559234	0.703470	Yes
F2-2	0.582879	***0.905079***	0.558851	0.713068	Yes
F2-3	0.668987	***0.993839***	0.529548	0.789173	Yes
F3-1	0.654699	0.422688	***0.728458***	0.667284	Yes
F3-2	0.663563	0.413298	***0.798689***	0.729609	Yes
F3-3	0.681142	0.366979	***0.880271***	0.735134	Yes
F4-1	0.702691	0.347238	0.563979	***0.913782***	Yes
F4-2	0.640825	0.365110	0.437679	***0.998270***	Yes
F4-3	0.637123	0.354688	0.520618	***0.994319***	Yes

1.5. Anticipated Results

For experimental data, we relied on the Superfamily 1.65 website (http://supfam.mrc-lmb.cam.ac.uk/SUPERFAMILY/cgi-bin/align.cgi). We pulled data for four protein families - acid proteases (F1), cytochrome b5 (F2), cytochrome-c (F3) and 4-helical cytokines (F4) - from the SCOP 1.63 Protein DataBase. The results of the experiment using a perceptron array are given in Table 3. (Sample F1-1 refers to sample #1 from family F1, sample F3-3 refers to sample #3 from family F3). We chose three random samples from each family (for a total of 12 samples), and we fed each sample into every network. Every sample scored the highest in its particular network. Although we obtained 100% correctness, however for certain cases, the difference between the highest and second highest score was not very significant (for example, results from the F3 family samples). With the multi-perceptron architecture, we hope to have a very significant difference between the highest and second highest score.

1.6. Future Work

There are thousands of protein families. It is impractical to use a single neural network to abstract every family. Our approach is to abstract one protein family with one neural network system. Our neural network will be based on either the perceptron, or the multi-perceptron. Once a network has been "tuned" to a particular protein family, when a sequence of that protein family is submitted to the network, that network will give a most positive output due to the resonance between the sequence and its family, relative to other networks (i.e., other protein families). This system is also useful for narrowing down the list of possible candidate families, in the event more than one network gives a high positive output.

References

1. Martin Chew Wooi Keat, Rosni Abdullah, Rosalina Abdul Salam: *Parallel Artificial Intelligence Hybrid Framework for Protein Classification* (2004). LSGRID 2004: 92-102.
2. Martin Chew Wooi Keat, Rosni Abdullah, Rosalina Abdul Salam, Aishah Abdul Latif: *Weightless Neural Network Array for Protein Classification* (2004). PDCAT 2004: 168-171.
3. http://library.thinkquest.org/18242/perceptron.shtml

4. Christopher Johansson, Anders Lansner: *A Parallel Implementation of a Bayesian Neural Network with Hypercolumns.* Department of Numerical Analysis and Computer Science, Royal Institute of Technology, Stockholm University, Sweden (2002). http://www.nada.kth.se/~cjo/
5. Jason Wang, Qicheng Ma, Dennis Shasha, Cathy Wu. *Application of Neural Networks to Biological Data Mining*. New Jersey Institute of Technology (2001). www.cis.njit.edu/~jason/
6. Jason Wang, Qicheng Ma, Dennis Shasha, Cathy Wu. *New Techniques for Extracting Features from Protein Sequences.* New Jersey Institute of Technology (2001). www.cis.njit.edu/~jason/
7. Laura Campitelli, Laura Delledonne, Alessandro Salvini. *A Neural Network Approach to Protein Sequence Processing*. Rome University, Italy (1998). http://kilab.csie.ncyu.edu.tw/course/machine%20learning/0900198.ppt

PROCESS INTEGRATION FOR BIO-MANUFACTURING GRID

ZHIQI SHEN*[1], HUI MIEN LEE[1], CHUNYAN MIAO[2], MEENA SAKHARKAR[3], ROBERT GAY[1] AND TIN WEE TAN[4]

[1]School of EEE, Nanyang Technological University, Singapore 639798;
[2]School of Computer Engineering, Nanyang Technological University, Singapore 639798;
[3]School of Mechanical and Aerospace Engineering, Nanyang Technological University, Singapore 639798 and [4]Department of Biochemistry, Yong Loo Lin School of Medicine, National University of Singapore, Singapore 117597
Email: Email: [1]{zqshen, eklgay, leeh0015}@ntu.edu.sg; [2]ascymiao@ntu.edu.sg; [3]mmeena@ntu.edu.sg and [4]tinwee@bic.nus.edu.sg

Recently there has been a great demand for extending high throughput life science research to bio-manufacturing. However, there is a gap from life science research to bio-manufacturing. Most existing bio-workflow tools/grid computing systems only provide an isolated solution to help bio-scientist to orchestrate bio-R&D operations such as bio-database query, bio-computation and analysis for biological problems in specific verticals. They are static and lack the ability to adapt in a dynamic changing environment. Manufacturing of biological materials such as diagnostics, therapeutics and prophylactics, or bio-manufacturing, typically involves many bioprocesses, each of which requires a set of bio-workflows to be choreographed. This is currently achieved by manually defining and managing the workflows for bioprocesses through different workflow tools. This paper proposes a novel goal-oriented approach to modeling bioprocesses, choreographing bio-workflows from different workflow tools, and to integrating agents, web services and workflows for automated execution. It demonstrates how a multi-agent system on a grid infrastructure can be further derived to adapt and automate complex bio-manufacturing workflow/processes in a dynamic changing environment. In this way, database access in a large data grids, high performance computing in a computational grid, and remote device control in a manufacturing grid, can be coupled to a supply chain management system to form a broad scale bio-manufacturing grid that streamlines the entire value chain from R&D, productisation and design of biological products.

1. INTRODUCTION

Recent outbreaks of highly contagious diseases worldwide highlight the urgent need to extend high throughput life science research to bio-manufacturing which aims to connect the current broad base of life science research to manufacturing areas and translate the technological know-how and research output into designs and subsequently manufactured products in a timely and rapid manner, and if possible, in quasi-real time.

Collaborative life science R&D activities revolve around data integration to computation integration and to virtual bio-labs with complex laboratory information management systems (LIMS). Bioscientists often need to perform experiments using shared data resources distributed worldwide. In addition to raw data resources that are shared on the web, many different bio-tools/applications that operate on it have also been developed, most of them with restricted functionality and targeted at performing highly specific tasks. A bio-R&D activity is performed through a set of bioprocesses. Collaborative life science R&D not only requires the sharing of bio-data resources but also the integration of bio-operations using various bio-tools. Therefore, it is both challenging and vital to integrate bio-tools/applications in various bioprocesses.

With the emergence of service-oriented technology, it is expected that any entity in interconnecting bio-applications/tools will be viewed as a service, whether it is a bio-instrument device data acquisition operation, or a bio-application/tool for signal processing or analysis of acquired data or a bio-database transaction. Web/grid services are fast emerging as enabling technologies for seamless bio-application integration. Workflows that orchestrate the bio-operations using different tools are poised to automate various bioprocesses to increase efficiency and productivity. As a result, many attempts for using web service technology and workflow management systems to tackle the above issues have been made [1-4].

However, most of existing bio-workflow tools/systems only provide an isolated solution to help bio-scientist to orchestrate bio-R&D operations such as bio-database query, bio-computation and analysis for specific bio-problems of limited scope. They are static and lack the ability to adapt in a dynamic changing environment. In the case of bio-manufacturing, many bioprocesses are involved, each of which requires a set of bio-workflows to be choreographed. This is currently achieved by manually defining and managing the workflows for bioprocesses through different workflow tools. They are also often not well interconnected with the research and development, or the discovery and design process.

Most of current research on integration of the available bio-services is based on web service architecture. For example, Taverna [1] (part of MyGrid project) is a leading research project of the UK government's e-Science programme. The Taverna software is a workflow workbench that provides a language and software tools to facilitate easy use of workflow and distributed services within the e-Science community.

With the growing number of bioinformatics resources such as computational tools and information repositories being made available as Web services, the Taverna project aims to provide a modeling tool with graphic user interface for designing and constructing bioinformatics workflows on top of bio-services over the web. Taverna provides user-friendly interfaces for bio-scientists to select and compose web services in a sequential order to form a workflow. Each operation in the defined workflow is able to invoke a specific bio-web service. Integrated bio-services for a specific problem are realized by running the corresponding workflows.

Some other research groups also focus on bioinformatics workflows on top of the available bio-services. They provide similar tools that allow users to compose and execute workflows, such as Pegasys [2], Wildfire [3] and BioWBI [4]. One of the common limitations in these is that most of them do not support dynamic and adaptive workflows. Moreover, none of the existing workflow systems supports the integration of workflows defined by different workflow tools.

The "Integrated Bio-laboratory Manufacturing and Services System" was a national project of Singapore in 2005, aimed at investigating how an "Integrated Workflow Infrastructure" for offering manufacturing and services involving the integration of bioinstrumentation, measurement systems and related databases as web services, could be developed. Concurrently, since 1997, the Bioinformatics Centre of the National University of Singapore and subsequently, its spin-off company, KOOPrime, a Singapore based company, has developed a suite of products and solutions, KOOPlatform, which addresses the needs of life sciences processes, specifically in the area of genomics and proteomics research [5]. This suite is known as "Workflows for Life Sciences" originally developed for a GlaxoWellcome-funded natural products drug discovery research centre's IT operations.

Although attempts have been made to offer descriptions by manually categorizing the services and sharing related workflows, these workflow management systems only provide a partial solution to the service integration. Existing Bio-workflow systems such as Pegasys, Taverna, Wildfire, BioWBI and KOOPlatform have limited user-centric modeling, abstracting, reasoning and automation capabilities. They are useful for composing partial, low-level processes, but lack the ability to integrate and automate a complete pipeline from R&D to manufacturing.

To meet the above challenges, we present Goal Net [6], a goal-oriented approach for modeling, integrating and automating bioprocesses as well as bio-workflows that are adaptive to dynamic changing environments. It also demonstrates how an agent-oriented system can be further derived to automate complex bio-manufacturing processes in a service-oriented environment.

The activities carried out according to a requirement are usually organized in groups of inter-related activities called processes that can be seen as a set of operations, rules and constraints specifying the steps that must be taken, and conditions that must be satisfied, in order to accomplish a given goal. This new methodology will lead to the design of an integrated workflow infrastructure to straddle across manufacturing and services involving integration of bioinstrumentation, measurement systems, related databases, software and web-based services.

2. IMPLEMENTATION

We have developed a prototype of bio-manufacturing system using the approach presented in this paper. In this system, we use Taverna and KOOPlatform as the bio-workflow systems. Web services that wrap different bio-services are orchestrated through the two workflow systems respectively. Goal Net is used to choreograph the workflows for modeling different bioprocesses (Figure 1).

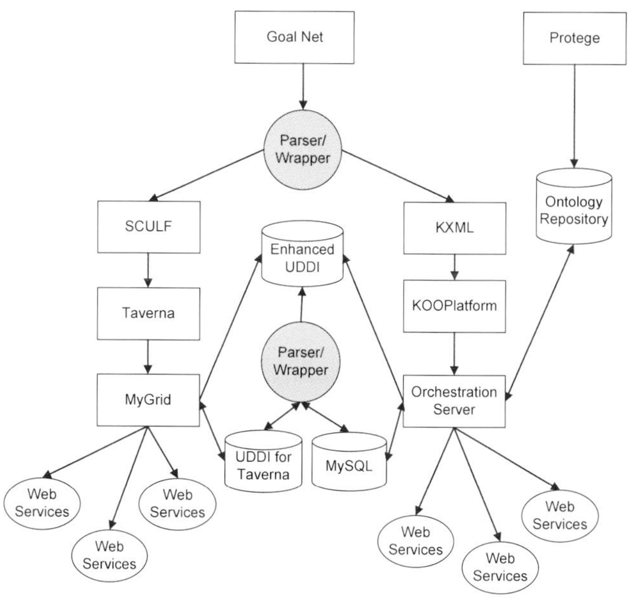

Figure 1. Architecture of the prototype system.

In this system, Taverna and KOOPlatform provide two sets of workflows. The extended UDDI provides common place for the services provided by the two systems. Goal Net provides a process integration platform for modeling bioprocesses and automating the bioprocess execution. The orchestration of existing bio-services not only needs a consistent definition of terminologies of bioprocesses but also the semantic linkage among various bioprocesses. The ontology repository is constructed in the system to store the defined consistent terminologies and concepts, and the semantics between the concepts.

The multi-agent development environment (MADE) that we have developed allows easy insertion of additional task libraries into the framework. The task libraries are the entry point where different functionalities can be added to MADE. To enable agents created by MADE to invoke web services and workflows in the two workflow systems, we have extended the MADE by adding two web service invocation components, Taverna integration component and KOOPlatform integration component.

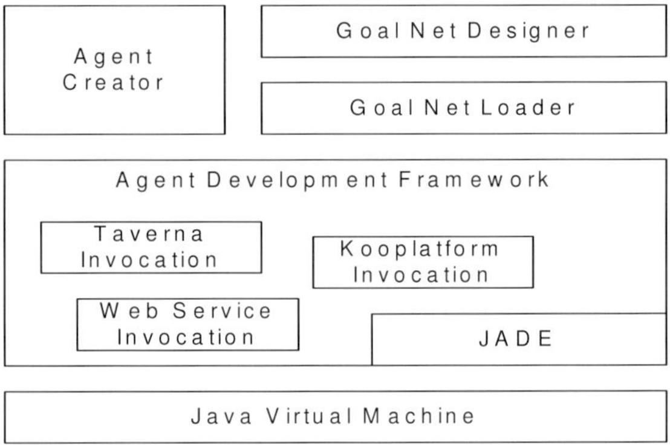

Figure 2. Structure of the extended MADE.

The Web service invocation component provides the API calls to the existing AXIS Web service tool [7] provided by Apache. AXIS is chosen because it is the latest Web service tool that provides good features with reasonable performance. By providing the API calls to the Taverna and KOOPlatform workflow systems, we have proven that the extended framework is able to act as a coordination framework for multiple atomic Web services and other existing Web services composition workflow models.

There are two types of storage methods for storing Goal Nets: XML-syntax description file and database storage. These two methods can be used to keep the goals, arcs, transitions, attributes and their interconnection information for dynamic Goal Net loading. As shown in Figure 3, there is a File/DB access library to access Goal Net storage system. The user can specify the rules and configuration for running agents through the user rule configuration file. This user rule configuration file will keep the information that agents require in decision making using action selection and goal selection inferences. JADE platform acts as agent creator which agents can be dynamically generated during run-time. Agents' deployment and un-deployment can be done through JADE services too.

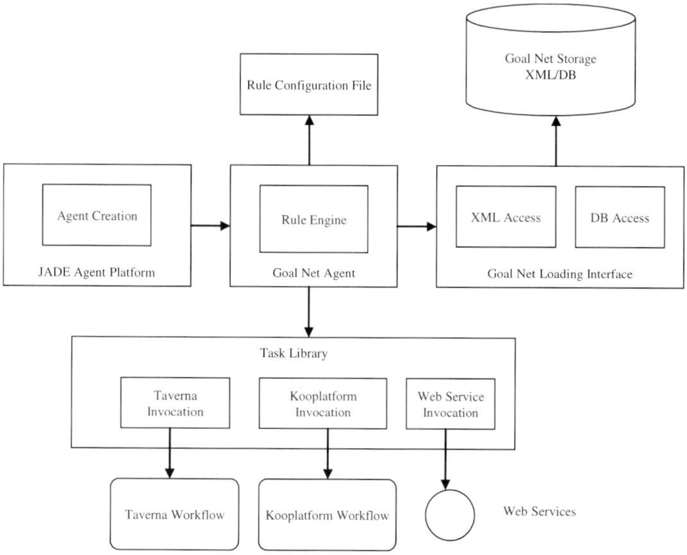

Figure 3. Agent-oriented architecture of the bioprocess execution system.

The action selection and goal selection mechanisms in the Goal Net model have been combined to become a rule engine. The Goal Net rule engine (Figure 3) makes decisions in goal selection and action selection based on the conditions defined in the user rule configuration file. Users need to specify the desired conditions and variables in the rule configuration file. Besides, users need to register tasks and the mapping of rules for the goal selection and action selection mechanisms to the rule engine. Finally web services and workflows can be invoked through the task libraries, that is, namely the web service invocation component, Taverna workflow invocation component and KOOPlatform workflow invocation component.

A typical scenario demonstrated by Taverna is that it provides a workflow to compare two genes X and Y. To illustrate our method, we designed a goal net to represent a process in which a ***compareXandY*** workflow will be invoked according to the user designed goal net. We then created an agent using the extended MADE and loaded the goal net to the agent. Figure 4 shows the running result of an invocation to a Taverna workflow from the agent.

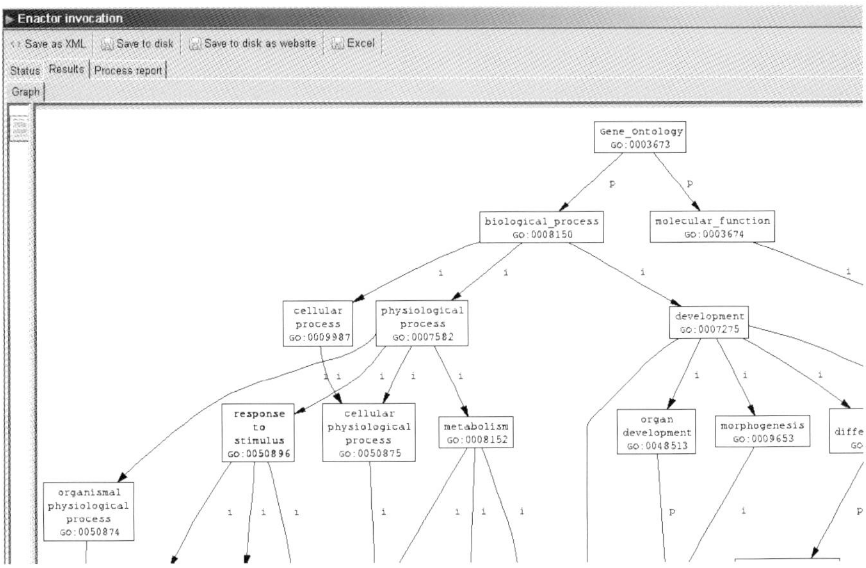

Figure 4. Goal Net agents can invoke Taverna workflows. The run results of the agent in Taverna are shown here.

Originally a user needed to prepare the data and invoke the workflows manually through the Taverna workbench, a GUI tool for Taverna workflow operations. With the system we have developed, previously designed and configured bioprocess can be stored in a database in the form of a goal net. In this way, a user only needs to create an agent and load the specific goal net to get the expected results and bypass the manual invocation. Furthermore the goal nets can be reused with different data for different requirements.

3. BIOPROCESS MODELING

A bioprocess is a specific ordering of activities with clearly identified inputs and outputs that achieve a certain goal. For example, a high level bio-manufacturing process takes a sample of an unknown infectious disease such as SARS or bird flu and produces a diagnostic material such as a DNA chip for rapid development of diagnostics and eventually for design of RNAi (miRNA/siRNA) therapeutics and DNA/peptide vaccines. The activities involved in a bioprocess can be bio-workflow executions, web service invocations, or other bio-application executions. The activities of a bioprocess and the order of the activities may be different in different situations. Currently most researchers in life science still manually manage the activities of a bioprocess, according to the current situation and based on their expertise, to adapt to the dynamic environment. In this paper, we adopted a goal-oriented approach to modeling bioprocesses by which life science researchers are able to transform their expertise to the process models. Then we build an agent oriented system to automate the process execution based on the process models.

A bioprocess model facilitates the alignment of bioprocess specifications with the technical framework that IT development needs. The challenges for modeling a bioprocess include:

1. The bioprocess model should capture relevant information consistently and thoroughly so that both life science researchers and the IT developers can understand the process requirements that are captured in the model.
2. The bioprocess model should capture alternatives and exceptions to standard in addition to normal operations.
3. The bioprocess model should be easily executed towards execution automation.

4. THE GOAL NET

In this paper, Goal Net is used to model the bio-manufacturing process. Goal Net is a composite goal hierarchy which is composed of goals and transitions. Round rectangles are used to represent the goals that agent need to go through in order to achieve its final goal. The transitions, represented by arc and rectangle, connect one goal to another specifying the relationship between the goals that it joins. Each transition must have at least one input goal and one output goal. Each transition is associated with a task list which defines the possible tasks that an agent may be required to perform in order to transit from the input goal to the output goal. Figure 5 shows a simple goal net.

Goal Nets can represent four types of basic temporal relationships between goals: sequence, concurrency, choice and synchronization. Sequence relationship represents a direct sequential relationship between one input goal and one output goal; concurrency relationship means one goal has more than one next goals, and all its next goals can be achieved simultaneously; choice relationship specifies a selective connection from one goal to other goals; synchronization relationship specifies a synchronization point from different input goals to a single next goal. With different combinations of the basic temporal relations, Goal Net supports a wide range of complicated temporal relations among goals. This is one of the major differences between Goal Net and other goal modeling methods.

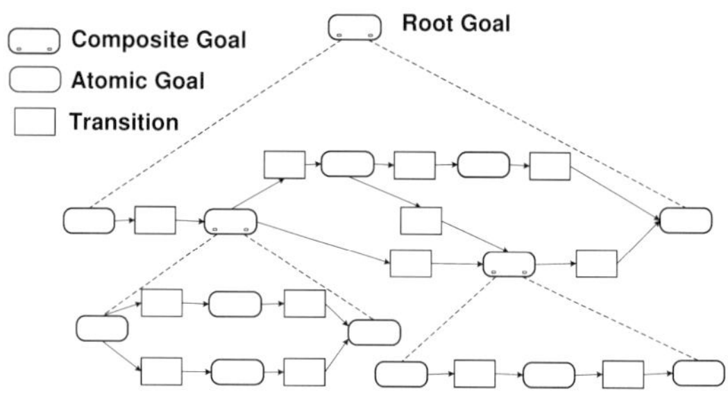

Figure 5. A simple Goal Net with two types of goals in Goal Nets, atomic goal and composite goal. An atomic goal accommodates a single goal which could not be split anymore; a composite goal may be split into sub-goals (either composite or atomic) connected via transitions.

In Goal Net, a composite goal needs to be decomposed into sub-goals. Here we do not intend to present a complete bio-manufacturing process model. Rather, we want to illustrate how a process can be mapped or modeled using Goal Net. A goal is a desired state that an agent intends to reach. Goal Net is an agent goal model. A goal net can be executed by an agent. We have proposed and developed an agent development framework based on Goal Net so that the created agent will refer to a goal net as its goal model to infer and guide its behaviors. When an agent is created, the agent has no goal and is running in an idle status, that is, it is an agent body. A goal net that represents a bioprocess is then loaded to the agent as its brain. Now the agent will start goal pursuit based on the goal net from the initial state to achieve the final goal of the bioprocess.

5. AUTOMATED BIOPROCESS EXECUTION

From an unknown infectious agent such as a deadly virus, to the elucidation of its complete genome, from its genome to the complete analysis and design of specific diagnostic DNA reagents, from the designed candidate diagnostics to the manufactured biochemical products, to the testing of these products against clinical samples, to the fine tuning and optimization of the diagnostic material such as a DNA chip – each of these steps can be individually semi-automated for high throughput today. Yet no-one has attempted to connect these disparate and distributed steps into a complete workflow chain of manufacturing and design steps end-to-end according to the best of our knowledge. Each step in the process can be modeled and choreographed on a software platform to achieve a specific high-level goal. The steps are then all represented digitally as services over a grid to be made available to and callable by geographically distributed life scientists through integrated workflow orchestration system as services. Together with these services, the relevant resources comprising human operator, machinery, equipment, laboratory instrumentation, materials and computers will also be made online over the grid. These steps will capture all the processes spanning over the entire value chain together with their relationships from business to operational to manufacturing. They are captured modularly and at different granularity. Furthermore, these steps are all semantically aligned and they are mapped ontologically and semantically for process integration and compatibility. Most of the steps are generic enough to be reconfigurable in different workflows.

In a bioprocess model represented by Goal Net, each step is to pursue a goal. An invocation of workflow system or a web service indicates a transition from one achieved goal to the next goal. Workflows and web services are invoked as tasks of the transitions in a goal net. In this way, the individual workflows and web services are integrated for bioprocesses by using Goal Net. The execution of a goal net represents the automated execution of the bioprocess represented by the goal net.

6. AGENT DEVELOPMENT ENVIRONMENT AND EXECUTION PLATFORM

Figure 2 shows the multi-agent development environment (MADE) and the execution platform we have developed. In this figure, Goal Net Designer is a tool to design bioprocesses using Goal Net. Agent Creator is built on top of the agent development framework [8] to provide an agent development environment.

Goal Net Loader is an interface for users to load a goal net to a created agent. Users can use it to assign different goal nets to different agents according to the users' requirements. In addition, MADE has been enhanced by incorporating the popular agent development environment JADE [9], which is in compliance with industry standard FIPA [10], supports standard agent communication mechanism and provides multi-agent running platform. Through the integration with JADE,

MADE can provide goal model development environment and at the same time, supports standard agent communication mechanism and agent running platform.

7. RESULTS AND DISCUSSION

With Goal Net, the composition of each bioprocess is designed in order to achieve a specific goal. A bioprocess can be decomposed into a hierarchy of sub-processes and activities. These sub-processes and activities are then assigned to different recognized workflows during run-time. Hence we see here a combination of various processes taking place at different locations of the virtual organizations in order to achieve the global goal of the high level process. The problem of the supervision or coordination of such a process at its various levels of decomposition is critical and especially so in this context, where the process and activities are not limited to a single organization, but to a set of autonomous, distributed, and heterogeneous nodes that need to cooperate. With Goal Net, the supervision and coordination are automatically derived during the process decomposition phase.

The advantages of using Goal Net include:

1. GoalNet is a novel goal oriented process modeling tool which can decompose a complex process (goal) into executable sub-processes (sub-goals) for achieving a common goal. The temporal relationships between processes (goals) are modeled. This is the key difference between Goal Net and other goal-oriented models.
2. GoalNet has reasoning capability. An agent running a goal net can reason the next goal to pursue and the next task to take for achieving the selected goal based on the current situation. Therefore the agent can compose the low level workflows to form a complete pipeline in a dynamic changing environment.
3. GoalNet is also a multi-agent modeling tool by which a multi-agent system can be derived from the process modeling for automating the processes execution.

Goal Net therefore provides a rich set of relationships and goal/action selection mechanisms to achieve a dynamic and highly autonomous process integration model. This approach of process integration is viable, in which:

1. The interactions between bioprocesses, bio-workflows and web services are represented as a Goal Net.

2. A bio-workflow or web service operation is represented by a transition task and a goal shows that a particular goal is reached after the execution of the transition tasks.
3. Combinations of different relationships between goals, sub-goals, and transitions can be used to represent complex bioprocess logic.
4. The dynamic bioprocess flow is achieved by defining action selection and goal selection mechanisms.

The goal-oriented bioprocess modeling proposed in this paper can handle atomic Web services and other existing web service compositions such as the workflows defined in Taverna and KOOPlatform. This is easily achieved through calling the APIs provided by the external composition models as transition tasks of Goal Nets.

With such a composite goal hierarchy and various temporal relations within the hierarchy, a complex system can be recursively decomposed into sub-goals and sub-goal-nets. In such a manner, a system can be easily modeled and simplified. For example, Figure 6 shows a bio-manufacturing process. Figure 7 shows the goal net which models the process.

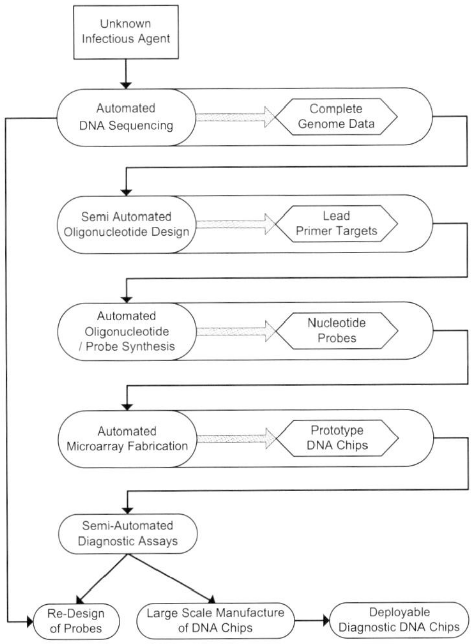

Figure 6. A bio-manufacturing workflow process with an underlying bio-manufacturing grid can be used to design, prototype and scale-up the manufacture of DNA chips against initially unknown infectious agents.

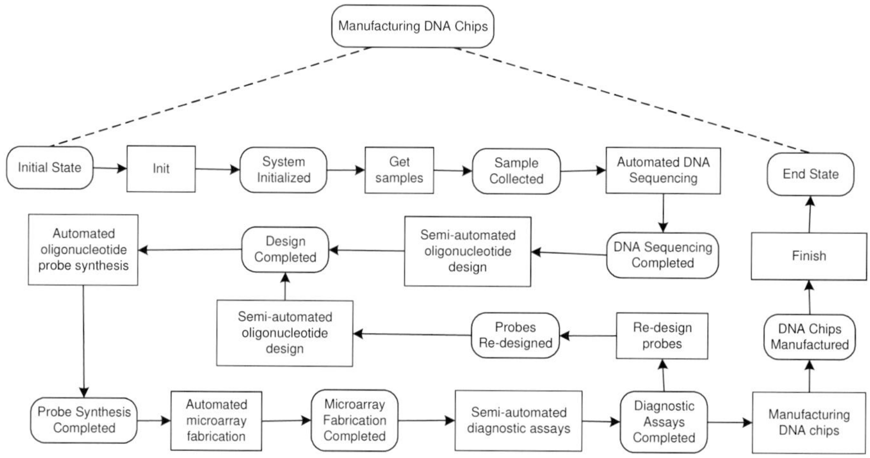

Figure 7. The corresponding Goal Net that models the bio-manufacturing workflow process in Figure 6.

8. CONCLUSION

To be able to integrate a complete pipeline from R&D to the manufacturing of diagnostic kits is not only an important advancement in terms of R&D value generated, but also vital in both economic and social context as it will allow us to build up a platform that can effectively respond against new outbreaks of infectious disease or bioterrorist attacks.

In this paper, we present a goal-oriented approach for bioprocess modeling and integration. A multi-agent platform for integrating bio-workflows and automating the bioprocess execution has been presented. The results generated by the developed prototype system shows that our goal to integrate various existing workflows and to automate the process execution using proposed approach has been achieved.

In fact, the core of biopharmaceutical manufacturing of PCR diagnostics, RNAi (RNA interference) agents, peptide vaccines etc, are processes which assemble linear polymers of biochemical monomers from a linear sequence of genetic information, mimicking the way each living cell does it. Automated DNA/RNA sequencers, oligonucleotide synthesizer machines, peptide synthesizers etc are readily available bio-instruments today with a plethora of service vendors on remote locations. Each instrument can be called to produce specific sequences by sending to the machines a text file generated from database searches and bioinformatics computation. Each system relies on standard sets of reagents and buffer solutions, which constitute the supply chain manufacturing and management system which, when integrated with these systems, will allow high throughput, automated or semi-automated manufacturing to take place.

Software Availability and Requirements

The source code and the executable tool are available at the site http://www.ntu.edu.sg/home/zqshen/imss-bio. The system requires Taverna 1.0, KOOPlatform 4.0, Java Runtime Environment 1.5 or higher.

ACKNOWLEDGEMENTS

We thank Kuay Chongthong and Xiong Luying for having developed the prototype system during their study for Masters degrees. We also thank Lim Teck Sin from KOOPrime Pte Ltd for his support during our research.

REFERENCES

[1] Oinn T, Addis MJ, Ferris J, Marvin DJ, Greenwood M, Carver T, Wipat A and Li P. Taverna, lessons in creating a workflow environment for the life sciences. *In* GGF10; Berlin, Germany. (2004)

[2] Shah SP, He DYM, Sawkins JN, Druce JC, Quon G, Lett D, Zheng GXY, Xu T, and Ouellette BFF. Pegasys: software for executing and integrating analyses of biological sequences. BMC Bioinformatics, **5**:40. (2004)

[3] Tang F, Chua CL, Ho LY, Lim YP, Issac P and Krishnan A. Wildfire: distributed, Grid-enabled workflow construction and execution. BMC Bioinformatics **6**:69. (2005)

[4] Leo P, Marinelli C, Pappadà G, Scioscia G and Zanchetta L. BioWBI: an Integrated Tool for building and executing Bioinformatic Analysis Workflows. In BITS2004; Mar 26-27 2004; Padova, Italy. (2004)

[5] KOOPlatform [http://www.kooprime.com]

[6] Shen ZQ, Gay R, Miao CY and Tao XH. Goal Oriented Modeling for Intelligent Software Agents. *In* IEEE/WIC/ACM International Conference on Intelligent Agent Technology (IAT'04); September 20 - 24, 2004, Beijing, China. (2004)

[7] WebServices - Axis [http://ws.apache.org/axis/]

[8] Shen ZQ, Gay R, Miao CY and Tao XY. Goal Autonomous Agent Architecture. *In* 28th Annual International Computer Software and Applications Conference (COMPSAC'04), September 28 - 30, 2004, Hong Kong, China. (2004)

[9] Bellifemine F, Poggi A and Rimassa G. JADE: a FIPA2000 compliant agent development environment. *In* 5th International Conference on Autonomous Agents, Montreal, Quebec, Canada. pp 216 - 217. (2001)

[10] FIPA Agent Management Specification [http://www.fipa.org/specs/fipa00023/]